BRILLIANT STARS

NOCTUA, CORVUS, CRATER, SEXTANS URANIÆ, HYDRA, FELIS.
Pl. 32.
Denebola
Regulus, vel Cor Leonis
CANCER
Vindemiatrix
Azimech, vel Spica
Zuben el Genubi
Zuben el Akrab
LIBRA
VIRGO
LEO
Al Chiba
Cor Hydræ vel Alphard
MONOCEROS
ATELIER TYPOGR
Sidy Hall, sculp.
LUPUS, CENTAURUS, ANTLIA PNEUMATICA, ARGO NAVIS, AND PYXIS NAUTICA.

BRILLIANT STARS

PATRICK MOORE

TED SMART

This edition produced 1997 for
The Book People Ltd
Hall Wood Avenue,
Haydock
St Helens
WA11 9UL

First published in Great Britain 1996 by Cassell
Wellington House, 125 Strand
London WC2R 0BB
in association with David Bateman Limited
30 Tarndale Grove, Bush Road, Albany
North Shore City, Auckland, New Zealand

British Library Cataloguing in Publication Data
A catalogue record for this book is available from the British Library

ISBN 1-8507-9322-0

Illustrations by Paul Doherty
Printed in Hong Kong by Colorcraft

Jacket illustration: Julian Baume (*Science Photo Library*)

CONTENTS

INTRODUCTION

Opposite Dawn with Venus, One Tree Hill, Auckland, New Zealand.
Photograph: John Dunlop, Auckland Observatory

Look into the sky on a dark, clear night, and you will see many hundreds of stars. They are not all alike; they are of different colours, and of course they differ in brightness. Here and there you will see a star which stands out because of its brilliance. Altogether there are twenty-one of these "first magnitude" stars — and each one has its points of special interest.

Our Sun is a normal star; indeed, astronomers relegate it to the status of a yellow dwarf. It appears so much more splendid than the rest only because it is much closer to us. Its distance is "only" 93,000,000 miles (150,000,000 km), or one astronomical unit; all the other stars are so remote that their distances are measured in millions of millions of miles. Indeed, they are so far away that our normal units of length, such as the mile or kilometre, become inconvenient. Instead we use the velocity of light, which flashes along at 186,000 miles (300,000 km) per second. In a year, light can cover almost 6 million million miles, or over 9 million million km, and that is what we term a light-year. The nearest of all stars beyond the Sun is over 4 light-years' away.

The Sun is the centre of the Solar System, and is attended by nine planets, of which the Earth comes third in order of distance. Like the Moon, the planets shine only by reflected sunlight, and though some of them (Venus, Mars, Jupiter and Saturn) look very bright, they are not nearly so important as they seem. Unlike the stars, they shift quickly in relation to each other. The stars are so remote that their individual or "proper" motions are too slight to be noticed with the naked eye even over periods of many lifetimes, so that to all intent and purposes the constellation patterns do not change.

Most people can recognize at least a few of the constellations, such as the Great Bear, Orion, and (in the southern hemisphere) the Southern Cross, but it is important to remember that the constellation names mean nothing at all. The stars are at very different distances from us, so that the stars in any particular group are not genuinely associated with each other; we are dealing with nothing more significant than line of sight effects. The names, too, are quite arbitrary. We happen to follow the pattern adopted by the Greeks, who named the patterns after mythological gods and heroes, or sometimes after everyday objects. If we had chosen, say, the ancient Chinese system, our star maps would have looked quite different, though needless to say the stars themselves would have been exactly the same. Today we recognize a grand total of 88 constellations; some are large and bright, while others are so small and obscure that they seem to have no claim to separate identity.

Many star catalogues have been drawn up. In 1603 a German amateur, Johann Bayer, allotted the stars Greek letters, so that each constellation had its set — logically beginning with the first Greek letter (Alpha) and continuing with Beta, Gamma, Delta and so on. Later the first English Astronomer Royal, John Flamsteed, gave the stars numbers. Individual names are, in general, used only for the very brightest stars; most of these names, though not all, are Arabic. Thus in Canis Major (The Great Dog) the leader, Sirius, is also known as Alpha Canis Majoris and as 9 Canis Majoris.

A star's apparent magnitude depends upon how bright it looks. The scale works in the manner of a golfer's handicap, with the most brilliant performers having the lowest values; thus a star of magnitude 1 is brighter than a star of magnitude 2, 2 is brighter than 3, and so on. The naked-eye limit is about magnitude 6, and modern telescopes can go down to below magnitude 25. Four stars actually have minus magnitudes; on this scale the brightest planet (Venus) is of magnitude -4.4 at its best, while the Sun's magnitude is almost -27.

Conventionally, the 21 brightest stars are classed as being of the first magnitude; they extend from Sirius (-1.46) to Regulus in the Lion (+1.35). They differ widely in luminosity; thus according to the Cambridge catalogue, which I follow here, Canopus in the constellation of the Keel is 200,000 times as luminous as the Sun, while Alpha Centauri, in the Centaur, is less than twice as powerful as the Sun. They also lie at different distances, ranging from just over 4 light-years for Alpha Centauri out to 1800 light-years for Deneb in Cygnus (the Swan). If some malevolent demon suddenly snatched Deneb out of the sky, we would not know for another 1800 years.

There is also the question of position, which is given by the star's declination. In the sky, declination corresponds to latitude on the Earth, and is defined as the angular distance north or south of the celestial equator. Most of the first-magnitude stars can be seen, at one time or another, from all inhabited countries, though Europeans lose the last six in our list and New Zealanders are deprived of the first three. Ten of our stars are in the northern hemisphere of the sky, and eleven in the southern.

I hope that these notes are helpful in providing a general background. So having "cleared the air", so to speak, let us begin our survey — beginning, appropriately with the brightest star of all, the magnificent Sirius.

The Constellations

Eighty-eight constellations are recognized by the International Astronomical Union. The Latin names are always used; after all, Latin is still the international language even though nobody actually speaks it nowadays. Here is the list:

Name	English name	1st-mag. star(s)
Andromeda	Andromeda	-
Antlia	The Air-pump	-
Apus	The Bird of Paradise	-
Aquarius	The Water-bearer	-
Aquila	The Eagle	Altair
Ara	The Altar	-
Aries	The Ram	-
Auriga	The Charioteer	Capella
Boötes	The Herdsman	Arcturus
Caelum	The Graving Tool	-
Camelopardalis	The Giraffe	-
Cancer	The Crab	-
Canes Venatici	The Hunting Dogs	-
Canis Major	The Great Dog	Sirius
Canis Minor	The Little Dog	Procyon
Capricornus	The Sea-goat	-
Carina	The Keel	Canopus
Cassiopeia	Cassiopeia	-
Centaurus	The Centaur	Alpha Centauri, Agena
Cepheus	Cepheus	-
Cetus	The Whale	-
Chamaeleon	The Chameleon	-
Circinus	The Compasses	-
Columba	The Dove	-
Coma Berenices	Berenice's Hair	-
Corona Australis	The Southern Crown	-
Corona Borealis	The Northern Crown	-
Corvus	The Crow	-
Crater	The Cup	-
Crux Australis	The Southern Cross	Acrux, Beta Crucis
Cygnus	The Swan	Deneb
Delphinus	The Dolphin	-
Dorado	The Swordfish	-
Draco	The Dragon	-
Equuleus	The Little Horse	-
Eridanus	The River	Achernar
Fornax	The Furnace	-
Gemini	The Twins	Pollux
Grus	The Crane	-
Hercules	Hercules	-
Horologium	The Clock	-

Hydra	The Watersnake	-
Hydrus	The Little Snake	-
Indus	The Indian	-
Lacerta	The Lizard	-
Leo	The Lion	Regulus
Leo Minor	The Little Lion	-
Lepus	The Hare	-
Libra	The Balance	-
Lupus	The Wolf	-
Lynx	The Lynx	-
Lyra	The Lyre	Vega
Mensa	The Table	-
Microscopium	The Microscope	-
Monoceros	The Unicorn	-
Musca Australis	The Southern Fly	-
Norma	The Rule	-
Octans	The Octant	-
Ophiuchus	The Serpent-bearer	-
Orion	Orion	Rigel, Betelgeux
Pavo	The Peacock	-
Pegasus	The Flying Horse	-
Perseus	Perseus	-
Phoenix	The Phoenix	-
Pictor	The Painter	-
Pisces	The Fishes	-
Piscis Australis	The Southern Fish	Fomalhaut
Puppis	The Poop	-
Pyxis	The Compass	-
Reticulum	The Net	-
Sagitta	The Arrow	-
Sagittarius	The Archer	-
Scorpius	The Scorpion	Antares
Sculptor	The Sculptor	-
Scutum	The Shield	-
Serpens	The Serpent	-
Sextans	The Sextant	-
Taurus	The Bull	Aldebaran
Telescopium	The Telescope	-
Triangulum	The Triangle	-
Triagulum Australe	The Southern Triangle	-
Tucana	The Toucan	-
Ursa Major	The Great Bear	-
Ursa Minor	The Little Bear	-
Vela	The Sails	-
Virgo	The Virgin	Spica
Volans	The Flying Fish	-
Vulpecula	The Fox	-

The First Magnitude Stars, in Order of Apparent Magnitude

Star		Magnitude
1. Sirius	Alpha Canis Majoris	-1.46
2. Canopus	Alpha Carinae	-0.72
3.	Alpha Centauri	-0.27
4. Arcturus	Alpha Boötis	-0.04
5. Vega	Alpha Lyrae	0.03
6. Capella	Alpha Aurigae	0.08
7. Rigel	Beta Orionis	0.12
8. Procyon	Alpha Canis Minoris	0.38
9. Achernar	Alpha Eridani	0.46
10. Betelgeux	Alpha Orionis	0.5 (variable)
11. Agena	Beta Centauri	0.61
12. Altair	Alpha Aquilae	0.77
13. Acrux	Alpha Crucis	0.83
14. Aldebaran	Alpha Tauri	0.85
15. Antares	Alpha Scorpii	0.96
16. Spica	Alpha Virginis	0.98
17. Pollux	Beta Geminorum	1.14
18. Fomalhaut	Alpha Piscis Australis	1.16
19. = Deneb	Alpha Cygni	1.25
19. = Beta Crucis		1.25
21. Regulus	Alpha Leonis	1.35

The First Magnitude Stars, in Order of Absolute Magnitude

Star	Absolute Magnitude
1. Canopus	-8.5
2. Deneb	-7.5
3. Rigel	-7.1
4. Betelgeux	-5.6 (variable)
5. Agena	-5.1
6. Beta Crucis	-5.0
7. Antares	-4.7
8. Acrux	-3.9
9. Spica	-3.6
10. Achernar	-1.6
11. Capella	-0.7
12. Regulus	-0.6
13. Aldebaran	-0.3
14. Arcturus	-0.2
15. Pollux	0.2
16. Vega	0.5
17. Sirius	1.4
18. Fomalhaut	2.0
19. Altair	2.2
20. Procyon	2.6
21. Alpha Centauri	4

Absolute magnitude is defined as the apparent magnitude which a star would have if it could be seen from a standard distance of 32.6 light-years, which is very roughly 200 million million miles (over 320 million million km). From this distance Canopus would cast strong shadows — but our Sun would appear as a dim star of magnitude 4.8. A very rough guide may be useful:

Absolute magnitude	**Luminosity, Sun = 1**
-8.5	200,000
-7	52,500
-6	21,000
-5	8300
-4	3300
-3	1300
-2	520
-1	200
0	83
+1	33
+2	13
+3	5
+4	2
+5	0.8
+10	0.008
+15	0.00008

The First Magnitude Stars, in Order of Distance

Star	**Distance, light-years**
1. Alpha Centauri	4
2. Sirius	9
3. Procyon	11
4. Altair	17
5. Fomalhaut	22
6. Vega	25
7 . = Pollux	36
7. = Arcturus	36
9. Capella	43
10. Aldebaran	68
11. = Achernar	85
11. =Regulus	85
13. Spica	260
14. Betelgeux	310
15. Antares	330
16. Acrux	360
17. Beta Crucis	425
18. Agena	460
19. Rigel	900
20. Canopus	1200
21. Deneb	1800

The First Magnitude Stars, in Order of Declination (from North)

Star	Declination
(North)	
1. Capella	+46
2. Deneb	45
3. Vega	39
4. Pollux	28
5. Arcturus	19
6. Aldebaran	17
7. Regulus	12
8. Altair	9
9. Betelgeux	7
10. Procyon	2
(South)	
11. Rigel	-8
12. Spica	11
13. Sirius	17
14. Antares	26
15. Fomalhaut	30
16. Canopus	53
17. Achernar	57
18. = Agena	60
18. = Beta Crucis	60
20. Alpha Centauri	61
21. Acrux	63

(These declinations are given in round numbers.)

Following pages *Rho Ophiuchi in the region of Antares.*
Royal Observatory, Edinburgh

The size of Sirius and the Sun compared.

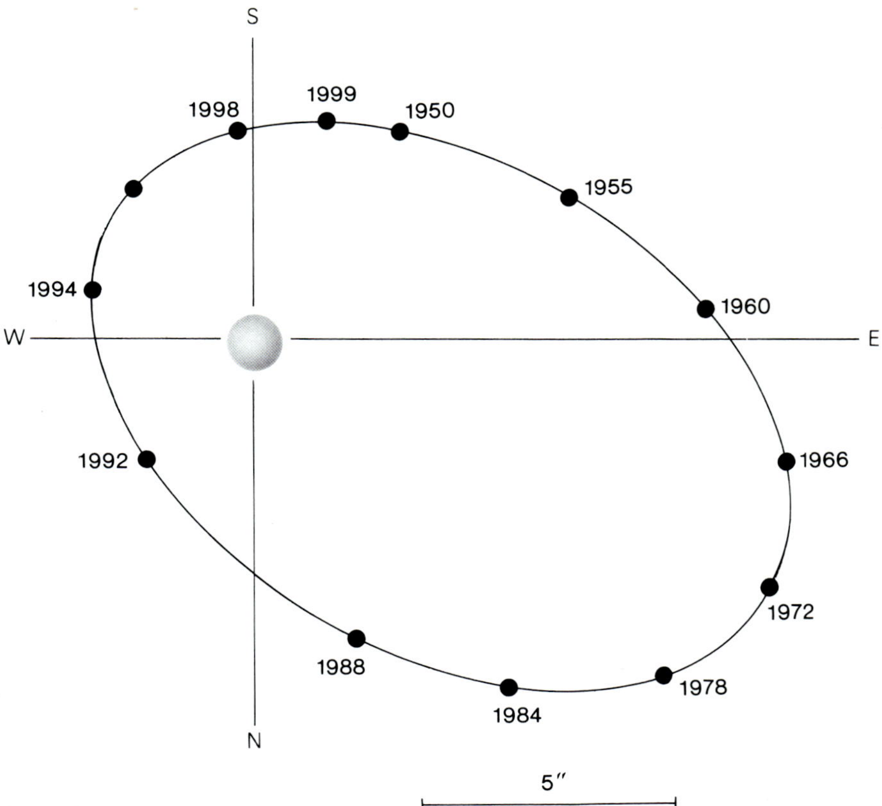

Apparent positions of Sirius B.

1. SIRIUS

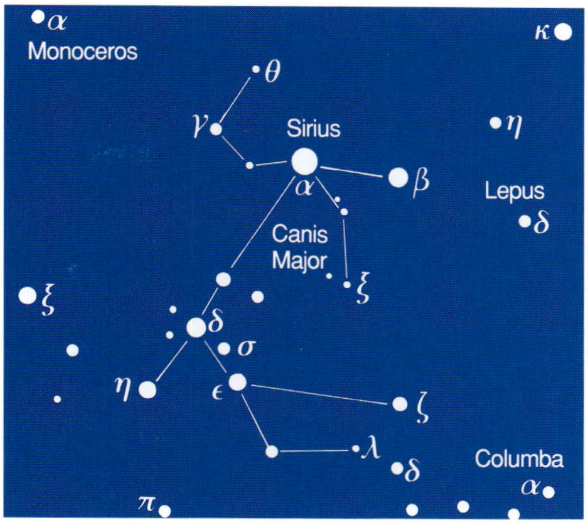

Canis Major (The Great Dog).

Alpha Canis Majoris: 9 Canis Majoris

Right ascension:	06h 4m 08s.9
Declination:	16° 42' 58"
Apparent magnitude:	-1.46
Absolute magnitude:	2.2
Spectral type:	A1
Luminosity, Sun=1:	26
Distance, light-yrs:	8.7
Parallax:	0".377
Radial velocity:	-8 km/s
Proper motion (per year):	RA -0".038, dec. -1".21

Of all the stars in the night sky, none can rival the glory of Sirius. It is three-quarters of a magnitude brighter than its nearest rival, Canopus, and almost a magnitude and a half brighter than Arcturus, the most brilliant star in the northern hemisphere of the sky. It is the leader of Canis Major, the Great Dog, part of Orion's retinue, and not unnaturally it is often called the Dog-Star.

Although Sirius is well south of the celestial equator, it is visible from almost all inhabited countries; even from places such as the Arctic towns of Tromsø in Norway or Fairbanks in Alaska it rises over the horizon, though admittedly not for very long. From anywhere south of latitude 74°S it remains on view all the time, provided that the sky is sufficiently dark and clear. Astronomers at the new South Pole Observatory see it throughout the six-months' night.

There is never any problem in finding Sirius, because it is so bright, but in case of any doubt simply follow through the line of the three stars of Orion's Belt. Of course, there are times when the stars in this area are too near the Sun to be seen, but Sirius is prominent for several months in every year. It is also easy to identify the other stars of Canis Major. They are not genuinely associated with each other, or with Sirius; in fact Sirius is much the closest of them, and lies in the foreground, so to speak.

The Greek Alphabet

In 1603 Johann Bayer introduced the system of allotting Greek letters to the stars; in a constellation the brightest star should be α (Alpha), the second brightest β (Beta) and so on, though in many cases the correct sequence has not been followed. Later John Flamsteed, the first Astronomer Royal, gave the stars numbers. Both systems are still in use, so that Sirius may be either Alpha Canis Majoris or 9 Canis Majoris.

The Greek Alphabet is:

α	Alpha	ν	Nu
β	Beta	ξ	Xi
γ	Gamma	ο	Omicron
δ	Delta	π	Pi
ε	Epsilon	ρ	Rho
ζ	Zeta	σ	Sigma
η	Eta	τ	Tau
θ	Theta	υ	Upsilon
ι	Iota	φ	Phi
κ	Kappa	χ	Chi
λ	Lambda	ψ	Psi
μ	Mu	ω	Omega

The first thing to notice about Sirius is that it twinkles. It also seems to flash various colours of the rainbow, though it is in fact a pure white star. Twinkling has nothing directly to do with the stars themselves, but is caused solely by the Earth's unsteady atmosphere, which "shakes" the starlight around before it reaches ground level. All stars twinkle, but the effect is particularly marked with Sirius, because it is so bright; it twinkles most when near the horizon — and from Britain, for example, it never rises very high. When a star is low down, its light has to pass though a greater thickness of air, and the twinkling is increased. From countries where Sirius can pass close to the zenith the effect is much less, though it is always perceptible.*

Naturally, Sirius has its place in ancient lore. Its name, unusually, is not Arabic, as with most other proper names of stars; it comes from the Greek, and is derived from σειριοσ, "sparkling" or "scorching". It really should be pronounced with a long i, and used to be written as "Syrius", though most people pronounce it "Sirrius".

To the poet Homer, Sirius was the Star of Autumn; Plutarch called it "the Leader", and the Greeks and Romans regarded it as unlucky. In Virgil's *Æneid* it is referred to as "the Dog Star . . . when he brings drought and diseases on sickly mortals, rises and saddens the sky with inauspicious light", though in 70 BC the Roman writer Geminus took a more reasonable view: "It is generally believed that Sirius produces the heat of the Dog Days, but this is an error, for the star merely marks a season of the year when the Sun's heat is greatest."

It was in Egypt that Sirius was regarded as vitally important. The Egyptian constellations were quite different from ours; there was no Great Dog, and on the famous Dendereh Zodiac, discovered by the French general Desaix in 1798 during Napoleon's adventures in Egypt, the stars in the area make up a cow lying on an Egyptian barque; the cow has a collar round its neck, with Sirius in between the horns. Sirius itself was worshipped as the Nile Star, Sothis, and its first appearance in the dawn sky every year — known as the helical rising — marked the annual flooding of the Nile, upon which the whole country's economy depended. It was less important to the Chinese, who called it Lang Hoo, but it was thought that when Sirius looked unusually bright it was time to beware of robbers and thieves!

Sirius is the closest of all the first-magnitude stars apart from Alpha Centauri. Its distance is 8.7 light-years, which works out at around 50 million million miles or 80 million million km; at the moment it is approaching us at 4½ miles (8 km) per second, though

* Planets twinkle less than stars, because a planet shows up as a small disk whereas a star is to all intents and purposes a point source. However, even a planet will twinkle when very low. The effect vanishes only if you go into space, above the top of unsteady air.

Old figure of the Great Dog.

this will not continue indefinitely and there is certainly no fear of an eventual collision. It was Edmond Halley, in 1718, who first realized that it has changed its position against the background of more distant stars by an appreciable amount since Ptolemy drew up his catalogue around AD 150; the shift since then has amounted to 1½ times the apparent diameter of the full moon. It is a member of what is called the Ursa Major moving cluster; other members are five of the stars in the Plough pattern, Menkarlina in Auriga, Rasalgethi in Ophiuchus, Zosma in Leo and Alphekka in Corona Borealis.

Rather surprisingly, Sirius is one of the least powerful of the first-magnitude stars; its eminence is due solely to its comparative nearness. It has a diameter of around 1,700,000 miles (2,700,000 km), and is "only" 26 times as powerful as the Sun. It pales by comparison with the other leaders of the Great Dog; thus Wezea or Delta Canis Majoris, with an apparent magnitude of 1.9, could match 130,000 Suns. Seen from our standard distance of 10 parsecs or 32.6 light-years, Sirius would be below the second magnitude, but Wezea would blaze down with an apparent magnitude of -8, so that it would cast strong shadows and would be conspicuous even in broad daylight. But Wezea is well over 3000 light-years away!

Edmond Halley. This is not Halley himself, but a waxworks model at Madame Tussaud's in London, where the author photographed him in 1996.

F.W. Bessel, the first man to announce the distance of a star beyond the Sun.

Sirius has a surface temperature of 10,000°C. It is 2½ times as massive as the Sun, and is using up its reserves of hydrogen "fuel" much more quickly, though it will take a long time to leave the Main Sequence and move into the giant branch of the H-R Diagram. Like all Main Sequence stars, it is officially classed as a dwarf.

In itself Sirius is unremarkable, but is not a solitary traveller in space; it has a companion, known officially as Sirius B and unofficially as the Pup, which has turned out to be a most extraordinary object.

When Friedrich Wilhelm Bessel had made the first measurement of the distance of a star — 61 Cygni, in the Swan — he turned his attention to Sirius, and made a careful series of measurements, using the large heliometer at the Konigsberg Observatory. The annual proper motion was not hard to detect, but to his surprise Bessel found something else. As Sirius tracked its way slowly across its background, it "weaved" its way along instead of moving in a straight line. Bessel realized that there could be only one explanation; Sirius was accompanied by a binary companion which was pulling the bright star out of position. From the weaving motion, Bessel concluded that the companion must be about as massive as the Sun, and that the orbital period was fifty years. He worked out where the companion ought to be, and did his best to find it, but despite all his efforts he could see no trace of it.

Bessel died in 1846, at the age of sixty-two. Shortly afterwards another talented mathematician, C.H.F. Peters, made similar calculations and came to the same result, but again the searches proved negative. The discovery was left to the American optical expert Alvan G. Clark, in January 1862. Clark — at that time much the best telescope-maker in the world — had just completed a new refractor with an 18½-inch object-glass, larger than any previously made. One

Very Luminous Stars of the Great Dog

Apart from Sirius, all the leaders of Canis Major are very hot, remote and luminous. They are, in order of apparent brightness:

Star	Proper Name	Magnitude.	Distance lt-yrs	Luminosity, Sun=1	Spectrum
α	Sirius	-1.5	8.6	26	A
ε	Adhara	1.5	490	5000	B
δ	Wezea	1.9	3060	130,000	F
β	Mirzam	2.0	720	7200	B
η	Aludra	2.4	2500	52,500	B
ζ	Phurad	3.0	290	420	B
o^2	-	3.0	2800	43,000	B

This means that Sirius has only 1/5000 the luminosity of Wezea!

Star Distances. If the Earth and Sun are represented by a 1-inch (3-cm) line, Alpha Centauri will be 4 miles (6 km) away.

of Clark's test objects was Sirius, because the great brilliancy would show up tiny faults in the optics which produce the dreaded telescopic "ghosts". There was indeed a speck of light near Sirius, but it was no ghost; it was the long-awaited companion, just about where Bessel and Peters had expected it to be.

In fact, the Pup is not particularly faint. It is of magnitude 8.6, and if it could be seen shining on its own it would be easy to see with binoculars, but it is so drowned by the glare from Sirius A that it is decidedly elusive even in a fair-sized telescope. The separation is never more than 11.5 seconds of arc, and when the separation is least, as in 1995, it is reduced to no more than 3 seconds of arc. In the 1970s I could observe it with my 15-inch reflector once I had blocked out the glare from the bright star by using an occulting bar in the optics of the telescope, but when I tried again in 1994, with the same equipment, I had no success. The orbit is decidedly eccentric (0.58), and the real separation between the two stars is on average 24 astronomical units, or well over 200,000,000 miles (320,000,000 km) — greater than that between our Sun and Mars.

The Pup has only 1/10,000 the luminosity of the Dog, and since its mass was known to be almost exactly the same as that of the Sun it was at first assumed to be large, cool and red. The situation was summed up in 1907 by J. Ellard Gore, an Irish astronomer who wrote some excellent popular books:

"If its faintness were merely due to its small size, its surface luminosity being equal to that of our Sun, the Sun's diameter would be the square root of 1000, or 31½ times the diameter of the faint star, in order to produce the observed difference in light. But on this hypothesis the Sun would have a volume 31,500 times the volume of the star, and, as the density of a body is inversely proportional to its volume, we should have the density of the Sirian satellite over 44,000 times that of water (the Sun's density being 1.5). This, of course, is entirely out of the question, and the result shows at once that the luminosity of the satellite's surface cannot possibly be comparable with that of the Sun. Its surface must be enormously less luminous than the Sun's surface."

Gore added: "The satellite has cooled down considerably, and is probably far advanced in regard to the total extinction of its light. It is unfortunate that its spectrum cannot be observed, as it should be a most interesting one."

In this latter statement Gore was certainly right — though not in the way he had expected. In 1915, the same year that Gore was killed in a street accident in Dublin, a spectrum was obtained by W. S. Adams, using the 60-inch reflector at the Mount Wilson Observa-

J.E. Gore

John Ellard Gore was one of the best-known Irish astronomers of the late 19th century — but he was never a professional astronomer. He was born at Athlone on 1 June 1845, and educated at Trinity College, Dublin; he became a civil engineer, and in 1868 went to India. In 1879 he retired from Government service, and devoted himself to astronomy. He discovered several variable stars (including the Mira star U Orionis) and computed the orbits of binary systems; he was also a skilled cosmologist, but was most celebrated for his popular books such as *The Visible Universe* (1893). He died in a street accident in Dublin.

tory. When Adams analyzed the spectrum, he had a shock. The Pup was not cool and red; its surface was hot — much hotter than the Sun's — and corresponded to a spectral type of F or early A.

If the Pup were both hot and faint, it had to be small. The diameter works out at about 26,000 miles (42,000 km), which is not much more than three times that of the Earth and is smaller than that of a planet such as Uranus or Neptune. This means that a vast amount of mass, over 300,000 times that of the Earth, has to be packed into a planet-sized globe, and the density rockets to an almost unbelievable level. Gore's estimate of 44,000 times that of water proved to be not "entirely out of the question", but a gross understatement. The real value is more like 125,000 times that of water. Sirius B, the Pup-star, is a white dwarf. The absolute magnitude is +11.4, and if we could bring a cubic inch of Pup material back to the Earth it would have a weight of about two and a half tons.

A white dwarf is a bankrupt star, lying to the lower left of the H-R diagram (see page 35). It has used up all its nuclear energy, and has no reserves left; it is shining feebly, but nothing but extinction awaits it, though the process of fading away to a cold, dead black dwarf takes so long that the universe may not yet be old enough for any black dwarfs to have evolved.

There are two mysteries associated with Sirius, one of which is worth taking seriously while, to be candid, the other is not. Taking the latter first, we may cast a brief look at the Dogon tribe in the African state of Mali. I have never been there, but neither has Robert K. Temple, a writer who published a book in 1975 claiming that the Dogon knew all about the Companion of Sirius, even to a good estimate of its 50-year orbital period. In the book it was also claimed that there were two French scientists who had learned, from the Dogon, that Sirius B is made of up "sagala", a material much denser than anything on Earth; it was added that the Dogon also knew about the rings of Saturn and the four large satellites of Jupiter.

Any theory, no matter how weird, is bound to draw support (remember the flying saucer craze; UFO societies still exist). But since in Dogon lore everything is "double", Sirius would not be expected to be the exception, and since French missionaries had been in Mali for some time, no doubt meddling in Dogan affairs, it is quite likely that astronomical matters would have been discussed.

The second mystery, which is quite different, concerns the colour of Sirius. Today it is pure white, but in ancient times it was often described as red — and this really is curious. The last great astronomer of the Classical period, Ptolemy, lived from around AD 120 to 180, and dwelt in Alexandria. His major work, which has come down to us by way of its Arab translation (the *Almagest*) is really a summary of all the scientific knowledge of the time, and it contains an excellent star catalogue. Ptolemy was certainly a skilled observer, and we still use the 48 constellations which he listed, including Canis Major. He listed six bright stars which were red: Arcturus, Aldebaran, Pollux, Antares, Betelgeux, and — Sirius. Of these Pollux is clearly "off-white", Arcturus light orange, Aldebaran and Betelgeux orange-red, and Antares very red. This leaves Sirius as the odd one out.

If this had been all, we would be tempted to say simply that Ptolemy made a mistake, but support comes from elsewhere. Homer, in the *Iliad*, compares Sirius' light with the gleam on the copper shield carried by Achilles; the poet Horace, around 65 BC, wrote that "the red Dog Star divides its children", and there are also significant comments by Pliny and Ovid. Seneca, who lived from 4 BC to AD 65, wrote that "the redness of the Dog Star is more burning; that of Mars is milder; Jupiter is not red" — and if Sirius were redder than Mars, its colour would have to be very pronounced indeed.

All this is very strange. At the present time there is absolutely no question about the whiteness of Sirius, and in the 10th century AD the Arab astronomer Al-Sûfi also called it white, so that if there has been any alteration it must have taken place well over a thousand years ago. Let us, then, look back over the story of the investigation.

Apparently the first near-modern writer to suggest that Sirius was once a red star was Thomas Barker, in 1760, who collected all the old references he could find and was convinced that there had been a colour-change. He was later supported by Baron Alexander von Humboldt, who wrote in his *Cosmos* (1845) that Sirius was "the one example of a star historically proved to have changed colour". Next came Dr. Thomas John Jefferson See (always remembered as T.J.J. See) who took up the case energetically, first in 1892 and then in 1927.

See, born in 1866, was admittedly an oddity. He came from Missouri, and worked successively at observatories in Berlin, Yerkes and Lowell before being appointed Professor of Mathematics in the U.S. Navy in 1899, subsequently becoming Director of the Naval Observatory. He carried out a great deal of work with respect to double stars and stellar evolution, and also observed the planets, claiming to have seen craters on the surface of Mercury long before they were recorded by the Mariner 10 spacecraft in 1973. It seems that he was a capable observer (though personally I have grave doubts about his Mercurian craters) and he was also a persistent historical researcher. On the other hand, he was universally disliked by all his colleagues, partly no doubt because he made a habit of claiming as his own discoveries which had been made by other people, but also because of what has been called his "repellent" personality. In 1898 he was dismissed from the Lowell Observatory at Flagstaff in Arizona, and the general opinion of him was aptly

White Dwarf to Black Dwarf?

Before a white dwarf loses the last of its energy, it must go through a stage when it has a cool surface. Yet energetic searches have failed to find any white dwarfs with surface temperatures of below 3000ºC, and this is why it is suggested that there might have been insufficient time for any black dwarfs to evolve. After all, the universe as we know it cannot be as much as 20,000 million years old, and recent researchers indicate that it may well be much younger than that. On the other hand, we cannot be sure — because a black dwarf, sending out no energy at all, will be very difficult to detect. We must wait and see.

Al-Sûfi and his Catalogue

The revival of astronomy at the end of the Dark Ages was due chiefly to the Arabs. The most valuable star catalogue was compiled by Al-Sûfi, who was actually a Persian nobleman; he lived from 903 to 986, and his catalogue includes 1018 stars, giving their magnitudes with considerable accuracy.

We know little about Al-Sûfi's life, but it seems that he was led to astronomy and mathematics by a love of his country's mythology and folklore.

summed up by one of his erstwhile colleagues, A.E. Douglass: "I have never had such aversion to a man or beast or reptile or anything disgusting as I have had to him. The moment he leaves town will be one of vast and intense relief, and I never want to see him again. If he comes back, I will have him kicked out of town." Perhaps fortunately, he did not return.

See spent a long time studying the Sirius problem. He analyzed the records left by Ptolemy and others, including the Roman writer Sextus Pompeius Festus, about whom not much is known. According to Festus, "ruddy dogs" were sacrificed at the Floralia festival in honour of the Dog Star, initiated in Rome in 238 BC following a decree by the oracle of the Sibyl. See commented: "Why ruddy dogs rather than dogs of any other colour?"

Certainly the problem is worth investigating, so let us look at the various possibilities. They are as follows:

(1). Sirius itself really has changed from red to white in historic times.

(2). It was the Pup which used to be red, and at that time outshone Sirius itself.

(3). Sirius was never red, but gave that impression because it flashes so strongly in various colours.

(4). There was formerly a cloud of material in interplanetary or interstellar space which lay between Sirius and ourselves, thereby reddening the star's light.

(5). The old reports have been misinterpreted, or were simply wrong.

Sirius is a perfectly stable Main Sequence star. It is not of the type to show rapid changes, and a switch from red to white over the past two thousand years would be very rapid indeed by cosmical standards. All in all, we can reject explanation (1) out of hand.

When we come to consider the Pup, we must be more cautious. Before a star becomes a white dwarf it must go through the red giant stage, and there is every reason to believe that the Pup did so. In that case it would have shone very brightly, and it would also have been very red. At first this seems to give a satisfactory explanation of the mystery — but closer examination shows that there are several fatal flaws in it.

The time-scale is all wrong. The change from red giant to white dwarf takes a great deal longer than 2000 years; 100,000 years would be a better estimate. True, a supernova outburst happens very quickly indeed, but neither component of Sirius is a supernova candidate; even the bright star is not nearly massive enough to explode in such a way. Also, it is not hard to work out the effect of combining the present Sirius with an even more luminous red companion. The two stars together would shine so brightly that they would be visible in

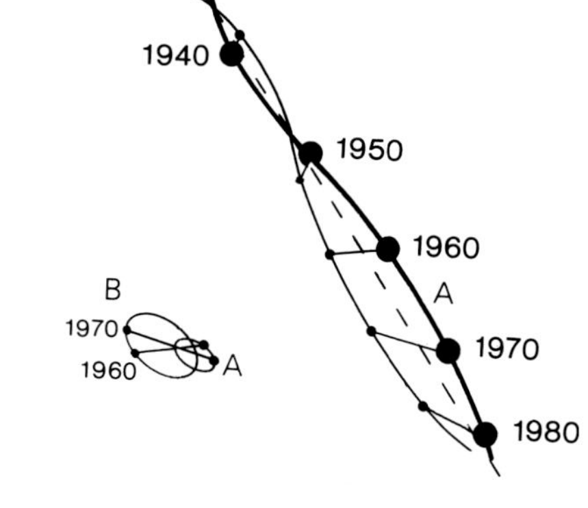

The motion of Sirius and Sirius B through the sky.

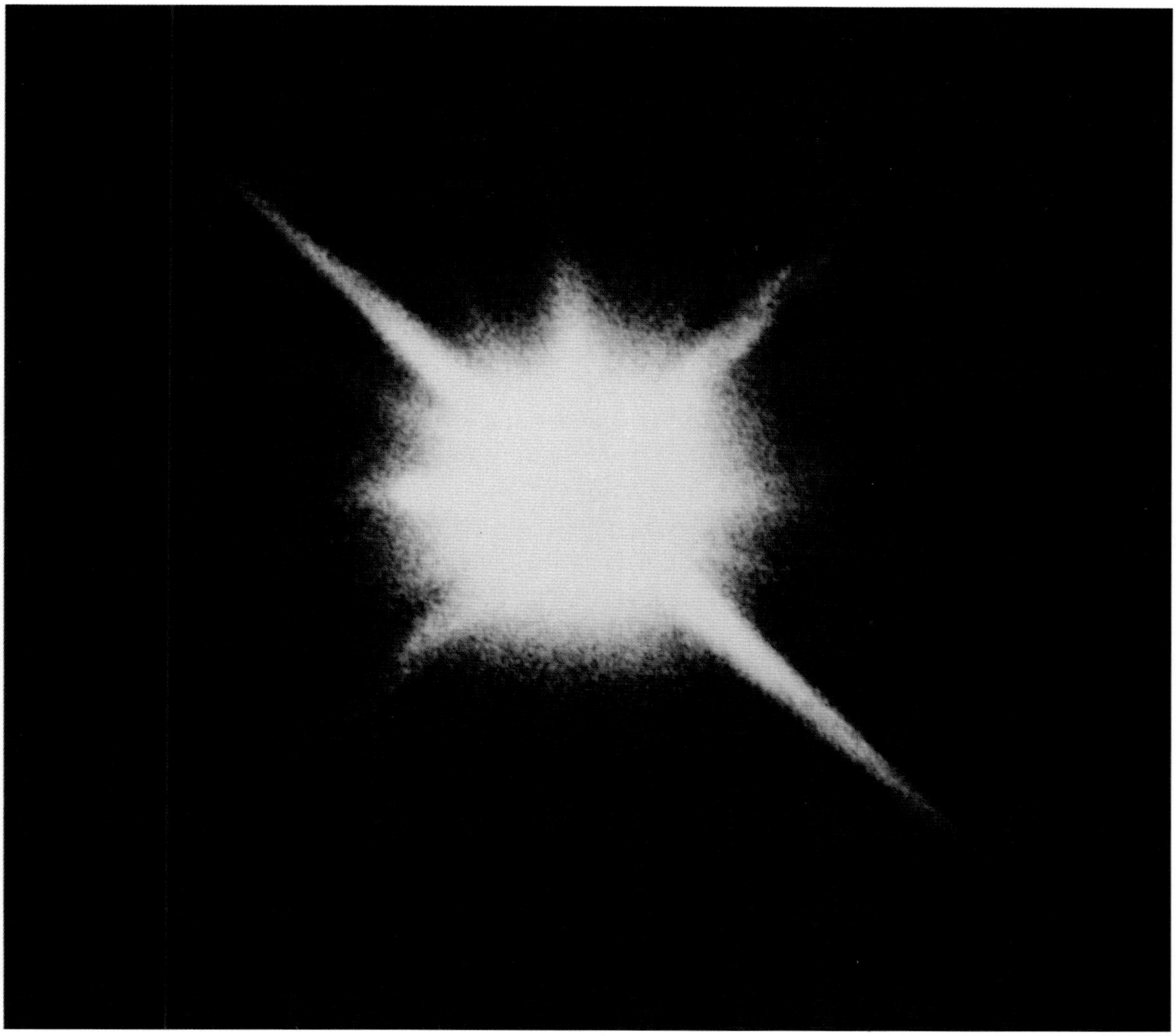

Sirius.
The spikes are, of course, optical effects.

daylight, and all the old observers would have commented on it. There have been suggestions that the Pup is surrounded by an unstable shell of hydrogen gas which can periodically but temporarily glow brightly, producing a red glare, but here too the total brightness of the system would be unacceptably great, and in any case careful spectroscopic work has revealed no trace of such a shell. It is equally improbable that the Pup itself is a close binary. So Explanation 2 must also be rejected.

I have already said something about the twinkling and flashing of Sirius, but redness is never dominant. Moreover, Ptolemy and the other old observers lived in regions from which Sirius rises to a great height above the horizon, and twinkling is much reduced. I have seen it straight overhead, and the scintillation is very slight, with no colour. I once carried out a test of my own, and in one of my television *Sky at Night* programmes on the BBC I asked viewers to look at Sirius on the next clear, dark night and let me know what they saw. I had over 5000 replies. Over 70 per cent recorded that the star was bluish or white; 14 per cent that it flashed all colours; 9 per cent that it was greenish-white; and only 4 per cent that it was yellowish or orange. Not one viewer saw Sirius as red, and the few who called it yellow or orange added that they had made their observations when the sky was somewhat misty, with Sirius low down.

It was Sir John Herschel who, in 1839, suggested the idea of a space-cloud which passed in front of Sirius and reddened its light as seen from Earth. This may be valid for the extraordinary variable Eta Carinae, and it was no doubt Eta Carinae which put the idea into Herschel's mind, but it does not seem at all plausible for Sirius, if only because we would almost certainly be able to detect signs of the cloud even now.

We seem therefore to be left with the final alternative. It is hard to credit some rather wild ideas (such as the claim that the old observers had defective colour vision — why should this affect only their view of Sirius?) and all things considered, it does seem that we have to dismiss all reports of a colour-change. Moreover, it is on record that in the first century BC a Chinese observer named Sima Qian described Sirius as white.

An element of doubt must remain, but the overwhelming opinion among astronomers — in which I fully share — is that the whole story of a change in Sirius must be dismissed as a red herring, or, perhaps more appropriately, a white herring.

What would we see if we lived on a planet in the Sirius system? Frankly, it seems most improbable that Sirius is a planetary centre; it is the wrong type of star, and from a planetary point of view the presence of the Pup would make the situation highly unstable. But if a planet did exist, its inhabitants would see two suns in their sky — one brilliant, one dim — and the view would be remarkable.

Meanwhile, Sirius shines gloriously down, and even with the naked eyes it is a lovely sight; use binoculars or a telescope, and it gives the impression of a flashing diamond. Sky-watchers of all nations enjoy it, even though it is no longer as important to us as it was to the Egyptians who watched for it at dawn as a sign that their life-giving Nile was about to flood.

2. CANOPUS

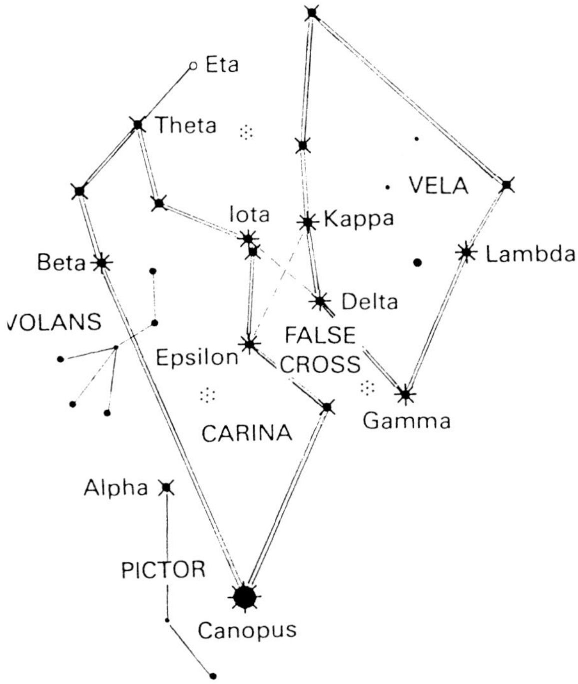

Diagram of Carina and Vela, parts of the old Argo Navis.

Alpha Carinae (formerly Alpha Argûs)

Right ascension:	06h 23m 57s.1
Declination:	-52° 41' 44"
Apparent magnitude:	-0.72
Absolute magnitude:	-8.5
Spectral type:	FO Ia
Luminosity, Sun=1:	200,000
Distance:	1200 light-years
Parallax:	0".028
Radial velocity:	+21 km/s
Proper motion (per year):	RA +0".003, dec. +0".02

I have always regarded Canopus as one of the loveliest stars in the sky. Unfortunately its position means that it cannot be seen from any part of Europe; it does not rise anywhere on Earth north of latitude 38°N. The fact that it can be seen from Alexandria, but not from Athens, was an early proof that the world is a globe; such behaviour could not possibly be explained on the theory that the Earth is a flat plane. It can be found by using Orion as a guide, but in fact this is generally unnecessary, because its great brilliance means that it stands out whenever it is above the horizon. It appears more than half a magnitude fainter than Sirius, but appearances are deceptive inasmuch as Canopus is much the more powerful of the two. When seen high in the sky it is truly magnificent; from most of New Zealand it is circumpolar, and over most of South Africa and Australia it sets only briefly.

Nowadays, Canopus is officially referred to as Alpha Carinae, leader of the Keel of the Ship, but in older lists it was still Alpha Argûs — and herein lies a tale.

It is glaringly obvious that the constellations are very unequal not only in brilliance, but also in size. It was Sir John Herschel, more than a century ago, who commented that the patterns seemed to have been drawn up so as to cause as much confusion and inconvenience as possible. Of course, a constellation pattern has no real meaning, because the stars lie at different distances from us, and we are dealing with nothing more significant than line of sight effects. One has only to look at the two Pointers to the Southern Cross, Alpha and Beta Centauri, about which I will have more to say in Chapter 3. They lie side by side, and make up a glorious pair, but there is no connection between them; Alpha is the nearest of all the bright stars, at little over 4 light-years from us, while Beta or Agena lies in the background, more than 450 light-years away.

All the old civilizations had their own constellation schemes, and if we happened to use, say, the Chinese or the Egyptian system, our maps would look very different from those we use, though the stars would be exactly the same. We follow the Greek pattern, and it seems to have been Aratus, in the third century BC, who first gave a reason for dividing the stars into groups; he said that it would be "impossible to know every single star or name them one by one". However, he was wrong. Only a few thousands of stars are visible with the naked eye, and learning them is not nearly so difficult as it might seem. Some enthusiastic astronomers — mainly amateur comet-hunters — go even further; they can identify the patterns at

Ptolemy. Whether he really looked like this we do not know — but he may well have done!

Ptolemy

Claudius Ptolemaeus (Ptolemy) apparently spent all his life in Alexandria; he was born about AD 120 and died around AD 180. About his personality we know absolutely nothing, but that he was a great scientist is not in dispute; periodical attempts to belittle him, and claim that he was a mere copyist, have been signally unsuccessful. He published a major work, the *Almagest*, which has come down to us by way of its Arab translation, and is really a compendium of all the scientific knowledge up to that time; its value to historians has been incalculable. It contains a star catalogue, which was based on earlier work by the Greek astronomer Hipparchus (c. 140 BC) but with important contributions from Ptolemy himself.

Ptolemy was a skilled observer, and he produced the first map of the civilized world which was based on more than mere guesswork. He believed that the Earth was the centre of the planetary system, and brought this theory to its highest degree of perfection, so that it is always called the Ptolemaic system.

Ptolemy's career marked the end of ancient astronomy — indeed, of ancient science. There followed a long period of stagnation, and little more was done until the rise of Arab science after the ninth century AD.

Canopus (in the centre of the picture) photographed by the author with a hand camera.

CANOPUS

SUN

The size of Canopus compared with the Sun.

Paul Doherty

An artist's impression of the Beta Pictoris system.
Paul Doherty.

least of thousands of telescopic stars, so that they can pick out any newcomer at once.

Ptolemy, last of the great astronomers of the ancient world, listed 48 constellations, and all of these are still in use, though in many cases with altered boundaries. Others have been added; Ptolemy spent all his life in Alexandria, so that stars of the far south were inaccessible to him, and some of the "modern" constellations have been given names which would have baffled Ptolemy; we have for example Telescopium (the Telescope) and Antlia (the Air-pump)! Others, introduced at various times, have been rejected, such as Officina Typographica (the Printing Press) and Lochium Funis (the Log Line). Modern maps recognise 88 constellations, and of these the largest is Hydra, the Watersnake, which covers 1303 square de-

The Voyage of the *Argo*

The story of the Argonauts is one of the most famous in Greek mythology. The ship was built to carry Jason and his companions on an expedition to secure the golden fleece of a ram which was in the possession of a powerful king, Aeetes, and was guarded by a particularly ferocious fire-breathing dragon. Jason was accompanied by many of the greatest Greek heroes — Hercules, Castor and Pollux, Orpheus and many others; the ship was named after its builder, Argus. Helped by Aeetes' daughter Medea, a powerful enchantress, the Argonauts were successful in seizing the golden fleece and taking it away.

grees of the sky even though it has only one brightish star — the orange Alphard, nicknamed the Solitary One because of its isolated position.

In the past times Hydra was surpassed by Argo Navis, the Ship Argo, named in honour of the vessel which carried Jason and his companions in their rather unprincipled quest for the Golden Fleece. Argo occupied well over 1800 square degrees, and contained a great many bright stars, so that it was hopelessly unwieldy. Eventually the International Astronomical Union, the controlling body of world astronomy, lost patience with it and chopped it up into a keel (Carina), a poop (Puppis), sails (Vela) and a compass (Pyxis, originally Malus, the Mast). In fact these divisions had been proposed much earlier by the French astronomer Lacaille, but it took a long time for the separate parts to become completely independent. Finally, the overall name of Argo was dropped.

This led to another complication. As we have seen, the leading stars in each constellation are allotted letters of the Greek alphabet; this was Bayer's system, introduced in 1603. With the dismemberment of Argo, the main stars were shared out according to their positions. Puppis took the old Zeta Argûs as well as Pi, Rho, Tau and others; Vela appropriated Gamma, Delta, Kappa, Lambda and Mu; the stars of Carina included Beta, Epsilon, Iota, Theta, Upsilon — and Alpha, or Canopus.

In Sumerian mythology — about which, admittedly, we do not know a great deal — it seems that the whole group of stars making up Argo was known by the name of Canopus. It may be that the name for the individual star was first used by the Greek philosopher Eratosthenes, around 230 BC; he called it Κανωποσ. Later this was turned into "Canobus", and then into the modern form of Canopus. Aratus called it "the Rudder", and wrote that

The slackened rudder has been placed beneath
The hind feet of the Dog.

The legendary Canopus was a character in the saga of the Trojan War; he was the pilot of the ship which carried King Menelaus home after the fall of Troy. When they landed on the coast of Egypt some way from Alexandria, Canopus had an unfortunate encounter with

The Areas of the Constellations

Now that Argo has been dismantled, the largest of the accepted 88 constellations is Hydra. The following are the only constellations with areas of over 1000 square degrees:

1.	Hydra	The Watersnake	1303 sq.deg.
2.	Virgo	The Virgin	1294
3.	Ursa Major	The Great Bear	1280
4.	Cetus	The Whale	1232
5.	Hercules	Hercules	1225
6.	Eridanus	The River	1138
7.	Pegasus	The Flying Horse	1121
8.	Draco	The Dragon	1083
9.	Centaurus	The Centaur	1060

Next come Aquarius, the Water-bearer (980), Ophiuchus, the Serpent-bearer (948), Leo, the Lion (947) and Boötes, the Herdsman (907).

The only constellations with areas below 100 square degrees are Circinus, the Compasses (93), Sagitta, the Arrow (80), Equuleus, The Little Horse (72) and Crux Australis, the Southern Cross (68). It may come as a surprise to find that the Southern Cross is the smallest constellation in the entire sky!

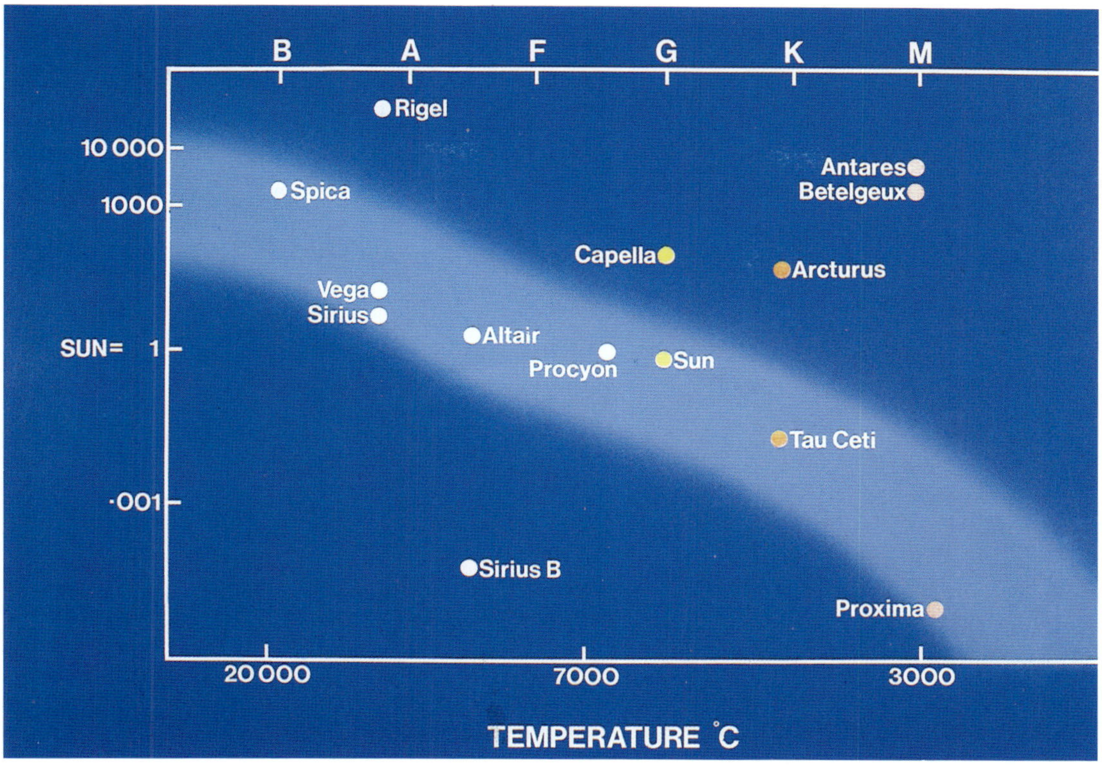

Typical Hertzsprung-Russell Diagram. In these diagrams, stars are plotted according to their luminosities and their spectral types.

a snake, and failed to survive. Menelaus had a monument built in his honour, and a town grew up around it. For a while the town of Canopus was quite important, but then it declined, and fell into ruins. Today the site is occupied by the city of Aboukir, which has come into near-modern history; it was near here, in 1798, that Napoleon's fleet was beaten by Nelson.

A legend related by the Arab astronomer Al-Sûfi, in the tenth century, is worth recalling. Canopus, or "Suhail", lay on the northern side of the Milky Way; he had two sisters, represented by the stars Sirius and Procyon. Canopus fell in love with Al Djauza (Rigel), but embraced her so clumsily that he killed her. To escape the wrath of his sisters, he fled south, over the Milky Way. Sirius followed him, but Procyon lacked the strength. This is why Sirius and Canopus are now in the southern hemisphere of the sky, while Procyon remains in the north.

Certainly the Arab nomads venerated Canopus, and used it for navigational purposes; they called it al-Fahl, the Camel Stallion. When it could be first seen rising in the dawn sky each year, it was a sign that summer had ended, and it was time for the weaning of camels. Thomas More, in his *Evenings in Greece*, wrote:

A camel slept — young as if wean'd,
When last the star Canopus rose.

Spectral Types of Stars

The main classes are as follows, though types R and N are now often combined as Type C.

Type	Colour	Surface temperature °C	Naked-eye example
W	Bluish	Up to 80,000	Regor (Gamma Velorum)
O	Bluish	40,000–35,000	-
B	Bluish-white	25,000–12,000	Spica
A	White	10,000–8,000	Sirius
F	Yellowish	7,500–6,000	Procyon
G	Yellow	6,000–5,000	Capella, the Sun
K	Orange	5,000–3,000	Arcturus
M	Orange-red	3,400–3,000	Betelgeux
R	Reddish	2,600	-
N	Reddish	2,500	-
S	Reddish	2,600	-

Many reddish and orange-red stars are variable in brightness. The system given here is that of Harvard, introduced by E.C. Pickering in 1890; originally it was meant to letter the classes A, B, C ... etc, but errors were introduced, and some classes were found to be unnecessary.

The Harvard system superseded an earlier system due to Angelo Secchi (1863-7), who divided the stars into four types:

I White or bluish stars (Sirius).
II Yellow stars (Capella, the Sun).
III Orange stars (Betelgeux)
IV Reddish stars, all below magnitude 5.

Canopus as an Aid to Space Navigation

Navigating a spacecraft is very different from navigating an aircraft. To determine the position of an unmanned vehicle far out in the Solar System, observations have to be made of celestial bodies. The Sun can, of course, be used, and another particularly useful object is Canopus, because it is so bright and is in a suitable position in the sky. In fact, the instruments used for this purpose are now usually called "Canopus sensors", though other stars can be used as well.

The Maori of New Zealand used it in a similar way; when it first appeared, certain crops were due to be planted. The Maori name for it was *Aotahi*. In Germany, later on, it was *Schif-stern*, the Ship Star.

Thomas Carlyle, in *On Heroes and Hero-Worship*, wrote that Canopus "shining down over the desert, with its blue diamond brightness (that wild, blue, spirit-like brightness far brighter than we ever witness here), would pierce into the heart of the wild Ishmaelite man, when it was guiding through the solitary waste there. To his wild heart, with all the feelings in it, with no *speech* for any feeling, it might seem a little eye, that Canopus, glancing out on him from the great, deep Eternity; revealing the inner splendour to him. Cannot we understand how these men *worshipped* Canopus; became what we call Sabeans, worshipping the stars?" Poetic indeed; but there is one point of special interest here, and this is connected with the colour of Canopus.

A star's colour depends on its surface temperature. Bluish stars are hotter than white, white hotter than yellow, yellow hotter than orange or red. Starlight can be analyzed by means of the spectroscope, and the stars have been divided into several well-marked spectral types, each denoted by a letter of the alphabet. In order of decreasing surface temperature, the types are: W (Bluish or greenish white), O and B (bluish), A (white), F (yellowish), G (yellow), K (orange), and M, R, N and S (orange-red). The sequence is alphabetically chaotic, but is easily remembered by the famous mnemonic "Oh Be A Fine Girl Kiss Me Right Now Sweetie", which is convenient even though it is no doubt Politically Incorrect. Canopus is type F, and therefore in theory should have a yellowish tinge, but I for one have never been able to see it as anything but pure white. The surface temperature is of the order of 7500°C, considerably greater than that of our yellow G-type Sun, whose surface is a little below 6000°C.

But Canopus is a very remote, very luminous star. According to the authoritative Cambridge catalogue, which I am following here, its luminosity is 200,000 times that of the Sun, or well over 7500 times greater than that of Sirius. Another way of expressing this is by absolute magnitude, defined as the apparent magnitude which a star would have if it could be seen from a standard distance of 32.6 light-years. (I will have more to say about this standard distance later.) From this range Canopus would shine as of magnitude -8.5, far brighter than Venus appears to us, and would cast strong shadows. Yet from the same distance our Sun would be a dim naked-eye object only just above magnitude 5.

However, there are reservations, because the distances of remote stars are by no means easy to measure with even reasonable accu-

Canopsus Road!

After the end of the war, a major observatory was set up outside Pretoria, in South Africa. The Radcliffe Observatory flourished for many years, and roads in the adjacent area were named after stars — Sirius Road, Rigel Road and so on. One of these was given the name of *Canopsus* Road. Jack Bennett, a prominent amateur astronomer, wrote to the Pretoria City Council and pointed out the error; it should be *Canopus*. The Council's reply was memorable. "We regret that all our maps have been printed, and the names cannot be changed now. Cannot you change the name of the star?"

The spread of Pretoria necessitated the closing-down of the Radcliffe Observatory, and the removal of all the instruments to a better site in Sutherland, Cape Province, and Canopus Road was re-named correctly!

Canopus Street, Pretoria.

The erractic Eta Carinae, imaged with the Hubble Space Telescope.

Eta Carinae

For a time, in April 1843, Canopus lost its status as the brightest star of the Keel. The title was temporarily usurped by Eta Carinae, the most erratic of all variable stars.

Eta was noted in 1677 by Edmond Halley, who made it magnitude 4. For the following century and a quarter it ranged between magnitudes 4 and 2, but in 1827 it suddenly brightened up to equality with Acrux, leader of the Southern Cross. In 1837 a new increase took place, so that Eta became as bright as Alpha Centauri. After a slight fading came the climax in April 1843; from all accounts Eta surpassed Canopus and almost equalled Sirius. Then, slowly, a decline set in. By 1870 the star had fallen below naked-eye visibility, and since then it has hovered around magnitude 7.

Telescopically, Eta Carinae has been described as "an orange blob" unlike a normal star; it is associated with nebulosity, and it is the strongest infra-red source in the sky beyond the Solar System. At its peak the luminosity may have been 6,000,000 times that of the Sun, and it is still immensely powerful, though much of its radiation now lies in the infra-red part of the spectrum. Its mass is at least 50 times that of the Sun, and it is highly evolved, so that its life-span is limited; eventually it will explode as a supernova. Recent observations made from the European Southern Observatory in Chile show that there are three companions.

Eta Carinae is unique, and there is always the chance that it will again flare up to dominate the whole of that part of the sky. Its distance is around 6400 light-years; it may well be the most luminous star known in our Galaxy.

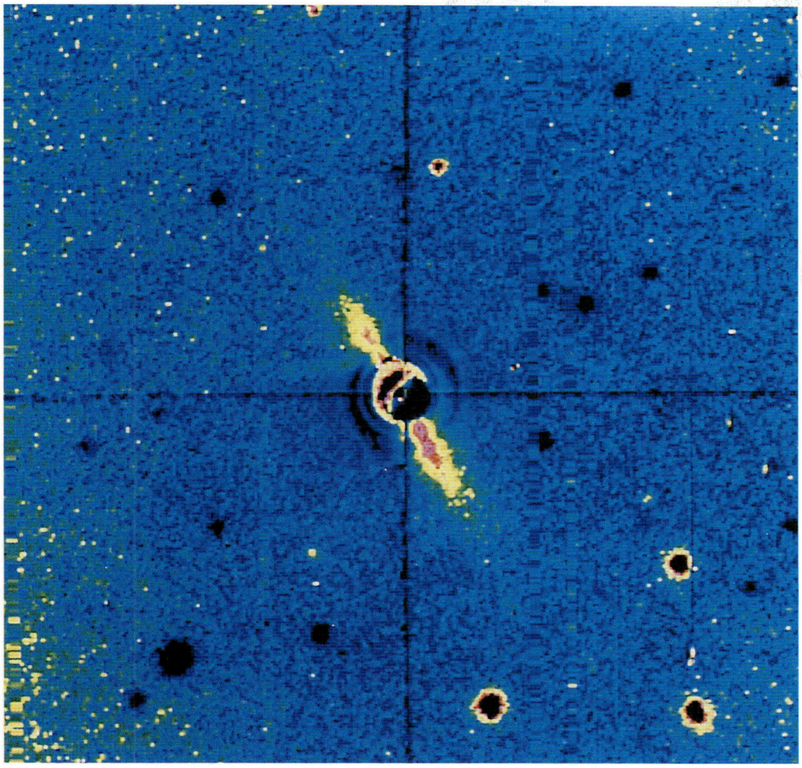

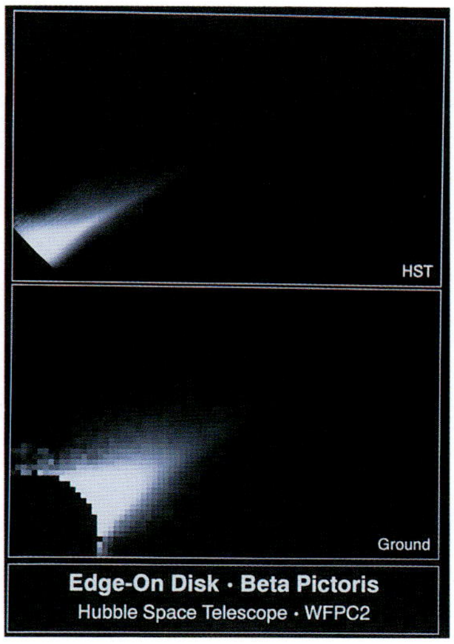

"Edge-on" view of the Beta Pictoris cloud; Hubble Space Telescope, 1996.

Left *The "cloud" round Beta Pictoris, as seen from the Las Campanas Observatory in Chile.*

racy. Their parallax shifts are very slight, and even a small error in measurement makes a great difference. With nearby stars, such as Sirius, we are entitled to be confident, but not so with Canopus. According to the Cambridge figure, its distance from us is 1120 light-years, but in another catalogue, that of K.E. Lang, the distance is a mere 98 light-years — in which case the absolute magnitude is -3.1, and the luminosity no more than 2000 times that of the Sun. Yet

Beta Pictoris

Adjoining Carina is the constellation of Pictor, the Painter (originally Equuleus Pictoris, the Painter's Easel). Near Canopus is the star Beta Pictoris, which is of magnitude 3.8; it is not conspicuous, and has never been given an individual name. It is of type A, 78 light-years away and 68 times as luminous as the Sun.

It was surveyed in 1983 by IRAS, the Infra-Red Astronomical Satellite, which operated for much of that year. Beta Pictoris was one of the stars found to be associated with a cloud of cool material which radiated at infra-red wavelengths; subsequently the cloud was detected optically by astronomers at the Las Campanas Observatory in Chile. It seems that Beta Pictoris is surrounded by a disk of material extending to nearly 50,000 million miles (80,000 million km) from the star; the disk is thin, and recent surveys by the Hubble Space Telescope confirm that there are "depleted regions" in it. It is possible, though by no means certain, that there is a system of planets, either in the process of forming or already in existence.

whichever estimate is correct, Canopus far outshines our own particular star, the Sun.

Star diameters are very hard to define. According to the best current estimate, Canopus is around 72,000,000 miles (115,500,000 km) across, which is large enough to contain the whole orbit of Venus round the Sun. It is, or course, very massive, and ranks as a supergiant, so that it is well advanced in its life-story; it has used up its original store of hydrogen "fuel", and has started to draw upon its reserves. Eventually, no doubt, it will explode as a supernova, and when this happens it will indeed be glorious, albeit for a brief period. Just when this will happen is uncertain — not for a very long time yet! — but happen eventually it must. In several thousand million years from now our Sun will still be shining, and will not be so very different from the Sun we know, but disaster will have overtaken Canopus.

It lies in a rich area. There are many bright stars round about,

The Milky Way near Eta Carinae. The Eta Carinae complex comes out well, even when taken with a hand camera.

such as Beta Carinae or Miaplacidus, not far below the first magnitude; two other stars of the Keel, Epsilon and Iota, make up the "False Cross" with Kappa and Delta Velorum. There are star-clusters and nebulae, some of which are very easy to find with binoculars, and in the little constellation of the Painter, adjoining Canopus, we find a fascinating star, Beta Pictoris, which may well be the centre of a system of planets.

Canopus dominates the whole of this region of the sky, but it has not always done so, because here too lies the extraordinary, erratic variable Eta Carinae, which is now below naked-eye visibility, but had its period of glory in the mid-nineteenth century, when for a while it outshone Canopus and almost rivalled Sirius. Whether it will ever return to this eminence remains to be seen; in any case, for the moment Canopus is supreme. It is a truly impressive star, a veritable cosmic lighthouse in our Galaxy.

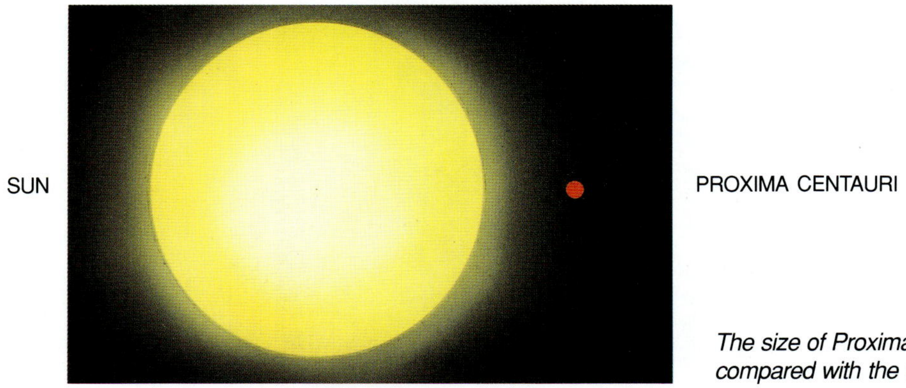

The size of Proxima Centauri compared with the Sun.

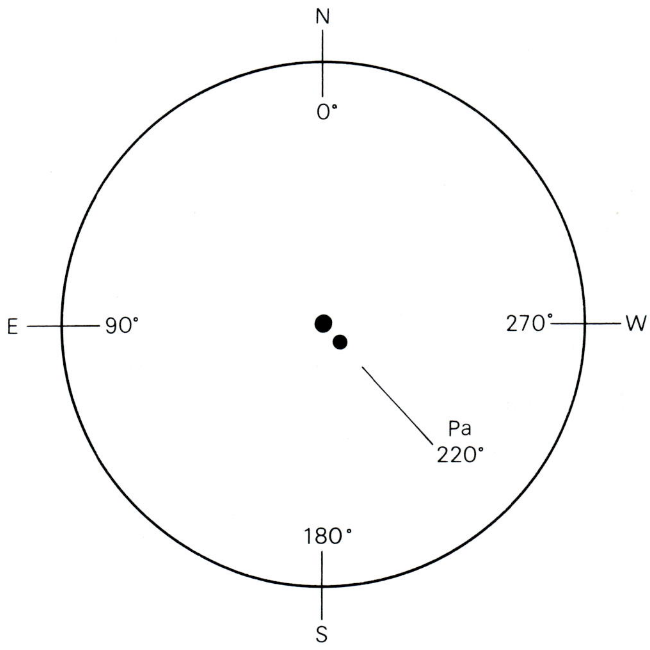

A position angle diagram of Alpha Centauri in 1997.

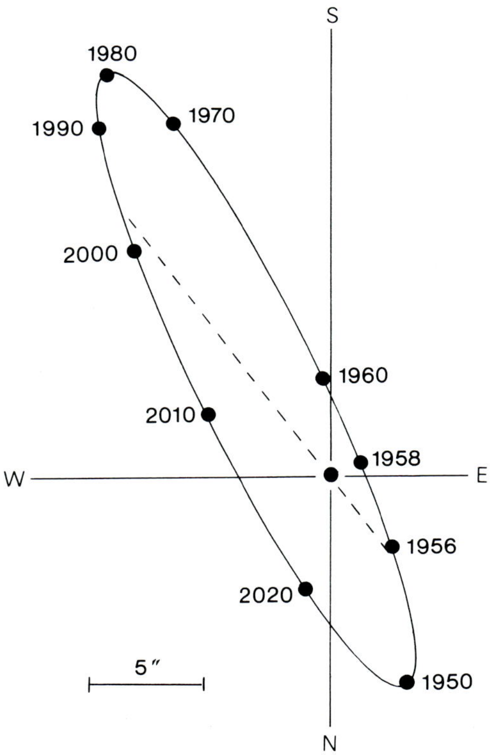

The apparent orbit of Alpha Centauri.

3. ALPHA CENTAURI

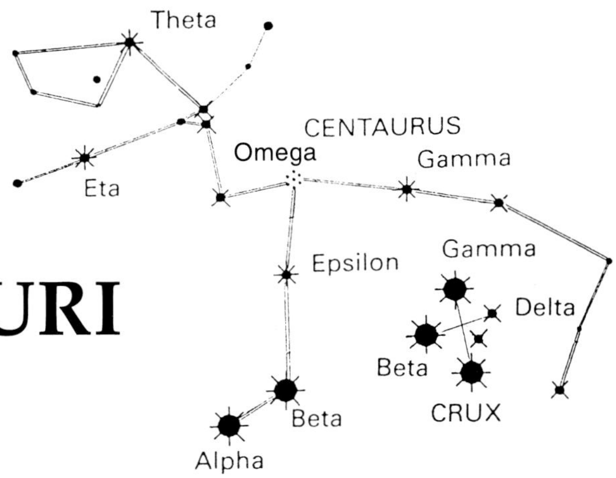

Centaurus and Crux.

	B (α^1)	A (α^2)	C (Proxima)
Right ascension:	14h 39m 35s.4	14h 39m 36s.7	14h 26m 18s
Declination:	-60° 50' 13"	-60° 50' 02"	62° 28'
Apparent magnitude:	1.39	0.00	11.0
Absolute magnitude:	+5.7	+4.4	+15.5
Spectral type:	K1	G2	dM5e
Luminosity, Sun=1:	0.4	1.7	0.00006
Distance, light-years:	4.4	4.4	4.3
Parallax:	0".750	0".750	0".772
Radial velocity:	-22 km/s	-22 km/s	22 km/s
Proper motion, RA (per year):	-0".494	-0".494	0".494
Proper motion, dec. (per year):	+0".69	+0".69	+0".69

Next in our list of "brilliant stars" comes Alpha Centauri, one of the two pointers of the Southern Cross. With the naked eye it appears as a single star of magnitude -0.27, considerably fainter than Canopus but appreciably brighter than its nearest rival, Arcturus. Yet Alpha Centauri is not a single star; it is a triple system, made up of two bright components and one very feeble dwarf. The nomenclature here is somewhat confusing. The brighter of the two main components is Alpha Centauri A, labelled Alpha^2; its magnitude is exactly zero. The secondary, Alpha Centauri B or Alpha^1, ranks as of magnitude 1.2. The dwarf, Alpha Centauri C, is puny indeed; you need a reasonably powerful telescope to see it at all.

Surprisingly, Alpha Centauri has never had a proper name which has been unreservedly accepted by astronomers. Bundula is sometimes found in old lists; also Toliman, the Hereafter. Ulugh Beigh, last of the great Arab astronomers, called it Rigil Kentauros, the Foot of the Centaur; to the Chinese it was Nan Mun, the Southern Gate; modern air navigators tend to refer to it as Rigil Kent. But to astronomers in general, it remains simply Alpha Centauri.

The Wise Centaur

In mythology, centaurs are generally described as wild, crude and lawless; Chiron, teacher of the Argonauts, was the exception. Half-horse and half-man, he was gentle, kind and wise. It was the Greek astronomer Eratosthenes, around 240 BC, who first identified Chiron with the constellation we now call Centaurus, and this certainly seems logical enough.

Centaurus itself may well commemorate Chiron, the half-horse, half-man teacher of the Argonauts; unlike most of the other centaurs, he was wise and kind. The association seems to be due to the Greek astronomer Eratosthenes, around 240 BC, and it is logical enough; after all, Chiron surely deserves a constellation!* Centaurus is large and rich; not much of it can be seen from Europe nowadays, though more of it was visible in ancient times (the effects of precession cause a small but remorseless shift in the position of the celestial pole). The declination of Alpha Centauri is -61 degrees, so that to see it you have to go south of latitude 30°N; thus it rises over Hawaii, but not over San Francisco. Of the leaders of Centaurus, only Theta, or Haratan, reaches a respectable altitude over any part of Europe. Haratan, just below the second magnitude, is one of our nearer neighbours in the Galaxy; it is 56 light-years away, and 17 times as luminous as the Sun.

In 1689 the Jesuit astronomer Father Richaud, observing from Pondicherry in India, was hunting for a comet when he discovered that Alpha Centauri is double. In 1752 it was reobserved by Nicolas Louis de Lacaille, who had sailed to the Cape to make observations of the far-southern stars, and from 1832 accurate micrometrical measurements of it were made by John Herschel, also at the Cape. It is a binary system — that is to say, the components are genuinely associated — and they move round their common centre of gravity in a period of 79.92 years. They were at their closest (periastron) in 1955, and their furthest apart (apastron) in 1995. The apparent separation ranges between 22 seconds of arc and 1.7 seconds of arc. The real separation is quite considerable, and the range is between 11 astronomical units and 35 astronomical units — one astronomical unit being the distance between the Earth and the Sun (93,000,000 miles or 150,000,000 km). This means that at periastron the distance between them is about the same as that between our Sun and Saturn, while at apastron it is considerably greater than the gulf between the Sun and Neptune. The orbital eccentricity is 0.5. From our point of view the orbit is highly inclined, and this makes it seem more elongated than it really is; however, Alpha Centauri is always an easy double, and virtually any small telescope will split it. In 1834 John Herschel called it "beyond any comparison, the finest double star in the sky".

*Chiron himself has had his name attached to an extraordinary member of the Solar System, discovered by Charles Kowal in 1977. It has been classed as an asteroid (No.2060) but moves well beyond the main asteroid belt, and spends most of its time between the orbits of Saturn and Uranus, completing one revolution in 51 years. Its diameter is around 150 miles, and it has developed a gaseous "fuzz" which has led some astronomers to re-classify it as a comet, though in this case it is a true cometary giant — even the nucleus of Halley's Comet has a longest diameter of less than ten miles. Other remote asteroidal bodies have since been found, and named after other centaurs such as Pholus. It may well be that they come from what is termed the Kuiper Belt, a whole belt of such objects lying not too far outside the orbit of Neptune.

The components are unequal in both brightness and colour. A (Alpha²) is 1.7 times as luminous as the Sun, with a similar spectrum, so that it is decidedly yellowish; B (Alpha¹) has less than half the Sun's luminosity, and is an orange star of spectral type K. However, B is the larger of the two. Its diameter is of the order of 1,100,000 miles (1,700,000 km) as against 925,000 miles (1,500,000 km) for A: the surface temperatures are respectively 5600°C and about 5000°C. A is 1.1 times as massive as the Sun, while B has 0.85 of the solar mass. Clearly, then, the masses are not so unequal as might be expected; smaller stars are always denser than larger ones.

Alpha Centauri is particularly important, from our point of view, because it is much the closest of all the bright stars. It is only 4.35 light-years away — about half as far as Sirius. It was also the first star to have its distance measured. The man who achieved this feat was Thomas Henderson, second Director of the Cape Observatory.

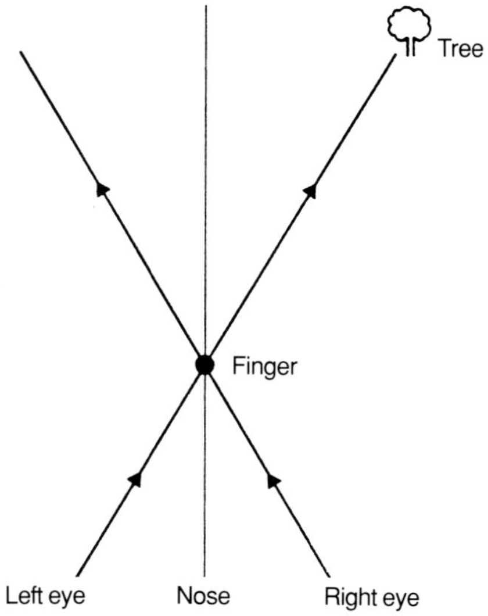

Principle of Parallax. Relative to the background tree, forefinger will show an angular shift when viewed with alternate eyes.

Henderson used the method of parallax, which is very easy to explain by means of a simple experiment, and is basically just the same as the method used by a surveyor to measure the distance of some inaccessible object such as a mountain-top. Close one eye, hold up a finger at arm's-length, and line it up with an object in the background, such as a tree in the garden. Now, without moving your finger or your head, use the other eye. Your finger will no longer be lined up with the tree, because you are observing it from a slightly different direction; your two eyes are not in the same place. If you know the length of the baseline (i.e. the distance between your eyes) and you measure the angular shift of your finger, which defines the parallax, you can solve the triangle by means of simple mathematics, and find out the real distance between your finger and your face.

Our surveyor needs a longer baseline, so he observes his mountain-top from two different observation points, suitably spaced out;

Position Angle and Separation

Position angle is defined as the apparent direction of one object with reference to another, measured from the north point of the main object (0 degrees) through east (090), south (180) and west (270) back to north. For Alpha Centauri, both the position angle and the separation alter quite quickly:

1970: position angle 204 degrees, separation 18".2.
1980: position angle 210 degrees, separation 21".8.
1990: position angle 215 degrees, separation 19".8.
2000: position angle 222 degrees, separation 14".2.

By November 2015 the separation will be reduced to 4 seconds of arc, after which it will start to increase again.

The globular cluster Omega Centauri (small-scale view).

Omega Centauri

Globular clusters are huge, symmetrical systems of stars, many with more than a million members each; they lie around the edges of the main Galaxy, and are very ancient, so that their brightest stars tend to be red giants and supergiants which are well advanced in their evolution. Over a hundred are known in our Galaxy, but only three are clearly visible with the naked eye: Omega Centauri and 47 Tucanae in the far south, and M.13 Herculis in the north. Of these the brightest, by far, is Omega Centauri, which looks like a hazy patch of about the fourth magnitude; in 1603 Johann Bayer regarded it as a nebulous star, and gave it a Greek letter. Its declination is -48 degrees, so that lies in the northern part of Centaurus and can be seen from anywhere on Earth south of latitude 42ºN; that is to say it just rises over San Francisco or Athens.

Its distance is about 17,000 light-years. Near its centre the average distance between its individual stars can hardly exceed a tenth of a light-year, so that to an inhabitant of a planet in that region there would be no night-time darkness at all, since many of the neighbouring stars would cast strong shadows. An Omega Centaurian astronomer could know little about the outer universe; if he could make out our Sun, it would shine as a very dim star of below the eighteenth magnitude!

Binoculars give superb views of Omega Centauri, and a small telescope is adequate to resolve the outer parts of the cluster into stars. Globular clusters are not unique to our Galaxy; other galaxies also are accompanied by systems of the same type.

he can then solve the triangle, as before, and obtain the distance of the mountain. But for a celestial body we need a longer baseline still, and we have to use more remote stars as reference points. This is what Henderson did. His baseline was the diameter of the Earth's orbit; 186,000,000 miles or 300,000,000 km. What he had to do was to measure the position of Alpha Centauri, against the background of remoter stars, over a six-months' interval. He found a slight but measurable parallax shift — 0.75 of a second of arc — and this gave him a distance of 4 light-years. Since one light-year is equivalent to rather less than six million million miles, it followed that Alpha Centauri is around 24 million million miles or 38 million million km away.

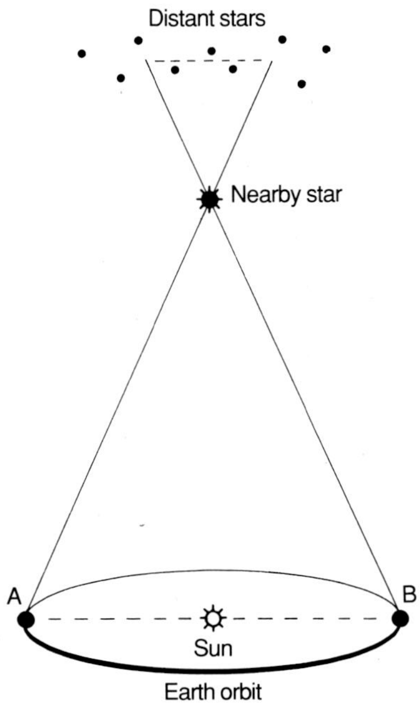

Parallax. The nearby star is observed from opposite sides of the Earth's orbit, and the parallax shift measured against the "background" stars, which are too far away to show detectable parallax shifts.

But Henderson had problems of his own, and although he had been appointed Director of the Cape Observatory he disliked the place intensely; he stayed for only thirteen months, and then retired thankfully to his native Scotland, departing in May 1833. He was in no hurry to work out his results, and needed some more observations — which of course he could not make from Scotland, because Alpha Centauri never rises there. It was not until 9 January 1839 that he finally published his findings, and by then he had been forestalled.

At the Königsberg Observatory in Germany, Friedrich Wilhelm Bessel had been working along the same lines, and he had concentrated upon 61 Cygni, a faint red dwarf pair of stars in the Swan. His reason for selecting 61 Cygni were the same as Henderson's for choosing Alpha Centauri; comparatively high proper motion, and the fact that 61 Cygni was a wide binary. Bessel began his work in 1838, and soon detected a parallax of just under 3 seconds of arc. His derived distance was 10 light-years, and as the real distance is 11.6 light-years he was very close to the mark. His announcement came in December 1838, a few weeks before Henderson's, and to Bessel, therefore, goes the honour of priority, even though Henderson's measures were actually made first.

At Dorpat, in what is now Estonia but was then part of the Russian empire, Otto Struve had been attempting to measure the parallax of the brilliant Vega. His results were much less accurate than those of Henderson or Bessel, which was understandable because Vega is much further away and shows a much smaller parallax shift, but his method was quite sound in theory, and it was not long before other stars too had their distances determined in the same way. Of course the method is limited, because beyond a distance if a few hundred light-years the parallax shifts become swamped in unavoidable errors of observation, but the main problem had been well and truly solved.

However, Alpha Centauri A and B are not the nearest stars

Proxima and Alpha Centauri.

beyond the Solar System. This distinction belongs to Proxima, otherwise known as Alpha Centauri C, the dim third member of the system.

Proxima was discovered in 1915 by R.T.A. Innes, also from South Africa; it lies almost two degrees from the AB pair, and is so faint that it is not easy to identify telescopically even when its position is known. The apparent magnitude is 11, and the absolute magnitude is below 15, so that it has only 0.00006 the luminosity of the Sun. It is orange-red, with an M-type spectrum; its distance from AB is around 13,000 astronomical units — that is to say, 400 times the distance between our Sun and Neptune. Probably it takes at least a million years to complete one revolution round AB, and at the moment its distance from us is 4.249 light-years.

Proxima is a true cosmic midget, perhaps 40,000 miles (65,000 km) in diameter, with a cool surface; if it could be seen from the distance of our Sun, it would produce only about 45 times the light of the full moon. It is what is termed a flare star, and undergoes sudden outbursts which make it brighten up perceptibly over a period of a few minutes, taking rather longer to subside back to normal. Undoubtedly this is due to localized activity on the star's surface, and indeed most stars, including the Sun, show flare activity; but an outburst on a powerful star passes unnoticed, whereas on a feeble dwarf it is very evident indeed.

But is Proxima a true member of the Alpha Centauri system at all? It has always been tacitly assumed to be so, but recently doubts have started to creep in, and there have been suggestions that it

Bessel and his Heliometer

Friedrich Wilhelm Bessel was born in 1784. He began work as an apprentice in a business firm, but soon turned to science, and in 1804 wrote his first astronomical paper, dealing with the orbit of Halley's Comet. The paper was published, and Bessel was introduced to Johann Schröter, an eminent amateur astronomer who had his private observatory at Lilienthal, near Bremen in Germany. Schröter engaged Bessel as his assistant, and he stayed for some years; in 1808 he moved to Königsberg to become director of the new observatory there. He remained at Königsberg for the rest of his life, and it was from here, in 1838, that he measured the parallax of 61 Cygni.

Bessel used a heliometer. This is a special form of telescope with an object-glass which has been cut in half, and therefore produces a double image of the object under study. The halves are then slid until the object appears single, and the amount of movement is a key to the actual apparent position of the object involved. Bessel was a superbly accurate observer; his heliometer was made by the best optical worker of the time, Josef Fraunhofer, and Bessel's results were of remarkable accuracy. Unfortunately he died in 1846, when still at the height of his powers.

R.T.A. Innes

Robert Thorburn Ayton Innes was born at Edinburgh, in Scotland, in 1861. He left school at the age of twelve, and thereafter was entirely self-taught. During his teenage years he showed exceptional mathematical ability, and at the unusually early age of seventeen he was elected a Fellow of the Royal Astronomical Society; but he had to earn his living — particularly after his marriage, in 1884 — and he emigrated to Australia, becoming a wine merchant. He borrowed a small telescope from a fellow amateur, W.F. Gale, and began to observe double stars. Before long he had discovered twenty new pairs, and sent his results to Sir David Gill, Director of the Cape Observatory in South Africa. Gill was impressed, and invited Innes to join the Observatory staff. Innes accepted without demur, and continued his double star work; by 1898 he had discovered another 280 new pairs.

In 1903 Gill was asked to nominate a director for the new Transvaal Observatory. It was meteorological, but Innes soon managed to introduce a programme of astronomy, and was able to persuade the Transvaal Government to purchase telescopes which were powerful enough to be really useful; eventually he obtained a 26½-inch refractor, which is still in use at Johannesburg and is known as the Innes Telescope.* Innes remained as Director until 1927, and carried out much important work; he also improved the design of the equipment, and undertook significant mathematical researches. He died in 1933.

Innes is probably best remembered for his discovery of Proxima, but certainly he ranks as one of the major astronomers of the early 20th century. He had no degree at all until 1923, when he received an honorary doctorate from the University of Leiden. He was unconventional in many ways. For example, he decided that the Transvaal heat made it uncomfortable to wear a tie — and thereafter he never did, even when being presented at the Dutch court!

*For a while, in the early 1990s, the telescope and its dome were commandeered by a curious society which allowed the dome to be used for social functions — and eventually they crashed the observing cage into the telescope, putting it out of action! I am glad to say that the situation has now been restored, and the telescope repaired; I like to think that I had something to do with this, as I was horrified when I found out what was happening, during a visit to Johannesburg in 1993, and raised the whole matter at an official level. The telescope is an excellent one, as I know; I used it in the 1960s and 1970s for mapping the Moon and observing Mars.

The "Dismal Swamp"

Thomas Henderson was born in Dundee, Scotland, in 1789. He built up a reputation as a fine astronomer, and seemed to be an excellent choice as the Director of the Cape Observatory, but he hated the place from the start, and in 1833 wrote to a colleague in the following terms:

"I will tell you about my residence in the Dismal Swamp among slaves and savages — plenty of insidious, venomous snakes. What would you think if, on putting out your candle to step into bed, you were to find one lurking behind the bed?"

He resigned in May 1833 without troubling to make more than a flimsy excuse, and was succeeded as Director by Thomas Maclear, who stayed until 1870 and built up the Observatory into a major astronomical centre. Henderson himself returned home, and became the first Astronomer Royal for Scotland.

Few pictures of Henderson exist. Apparently he had a pronounced squint and was reluctant to have his portrait painted.

The Innes Telescope at Johannesburg in January 1996 after it had been rescued and repaired.
Photograph: Patrick Moore

may be merely passing by at a speed too great for it to be held gravitationally by the bright pair. There is some extra evidence in favour of this idea. Our Sun must be of the order of 6 million million years old, and the bright components of Alpha Centauri are thought to be somewhat older than this. Flare stars such as Proxima are not,

The Nearest Bright Stars

Only eleven stars of apparent magnitude 2 or brighter are within 50 light-years of us. They are:

Alpha Centauri	4.35 light-yrs
Sirius	8.8
Procyon	11.4
Altair	16.6
Fomalhaut	22
Vega	26
Arcturus	36
Pollux	36
Capella	43
Castor	46
Menkarlina (Beta Aurigae)	46

For stars above apparent magnitude 4, we find thirty within fifty light-years of us: Epsilon Eridani, 10.7. Delta Pavonis and Eta Cassiopeiae, 19. Beta Hydri, 21. Chi Draconis and Xi Ursae Majoris, 25. Gamma Leporis and Mu Herculis, 26. Delta Eridani, 29. Zeta Herculis, 31. Eta Boötis, 32. Beta Trianguli Australe and Beta Virginis, 33. Alpha Hydri, Gamma Virginis and Beta Aquilae, 36. Beta Leonis, Iota Pegasi and Gamma Serpentis, 39. Beta Cassiopeiae, 42. Gamma Apodis, Beta Arietis, Alpha Circini, Theta Centauri, Theta Ursae Majoris and Delta Capricorni, 49.

Some of these stars, notably Delta Pavonis, are remarkably similar to the Sun. Whether or not they are centres of planetary systems remains to be seen.

Flying Stars

61 Cygni, the first star whose distance was announced, is known as the Flying Star because of its unusually large proper motion. It is not, however, the holder of the speed record. This is held by a dim red dwarf in Ophiuchus, studied by the American astronomer Edward Emerson Barnard and always known as Barnard's Star: the annual proper motion is over ten seconds of arc, but even so it takes around 190 years to cover a distance in the sky equal to the apparent diameter of the full moon. The brightest stars with really large proper motions are as follows:
Barnard's Star, annual proper motion 10.27 seconds of arc.

Lalande 21185	4.75
61 Cygni	4.12
Epsilon Indi	3.93
Wolf 359	3.84
Proxima	3.71
Alpha Centauri	3.61

Apart from Alpha Centauri, all these are relatively close-by red dwarfs.

in general, anything like as ancient; when a dwarf becomes senile, it loses its flare activity — and if Proxima has not yet reached this stage in its evolution, it is presumably considerably younger than A and B. We have to admit that we cannot be certain either way.

Because the Alpha Centauri system is so close, on the stellar scale, it would be fascinating to believe that there might be inhabited planets there, but obviously there are problems. When closest together, A and B are about as widely separated as the Sun and Saturn. Theoretically, then, either might be attended by close-in planets similar to our own Mercury, Venus, Earth or Mars, but one has the feeling that the presence of two suns would lead to a rather unstable situation and an erratic climate. Certainly any inhabitants of a planet in the system would have an interesting sky. From a planet moving round A, B would range between magnitude -18.1 and -20.6; from a planet associated with B, A would range between magnitude -19.4 and -21.9 (remembering that as seen from Earth, our Sun has an apparent magnitude of -26.8). The constellation patterns would not be so very different from ours; the Sun would show up as a reasonably bright star on the boundary of the constellation Cassiopeia.

European astronomers always regret that Alpha Centauri lies so far south in the sky, but from places such as Sydney, Cape Town and the whole of New Zealand it is circumpolar, while from the northern parts of Australia and South Africa it sets only briefly. Through any telescope it is a glorious sight, and it is hard to disagree with John Herschel's view that it is the most imposing of all double stars.

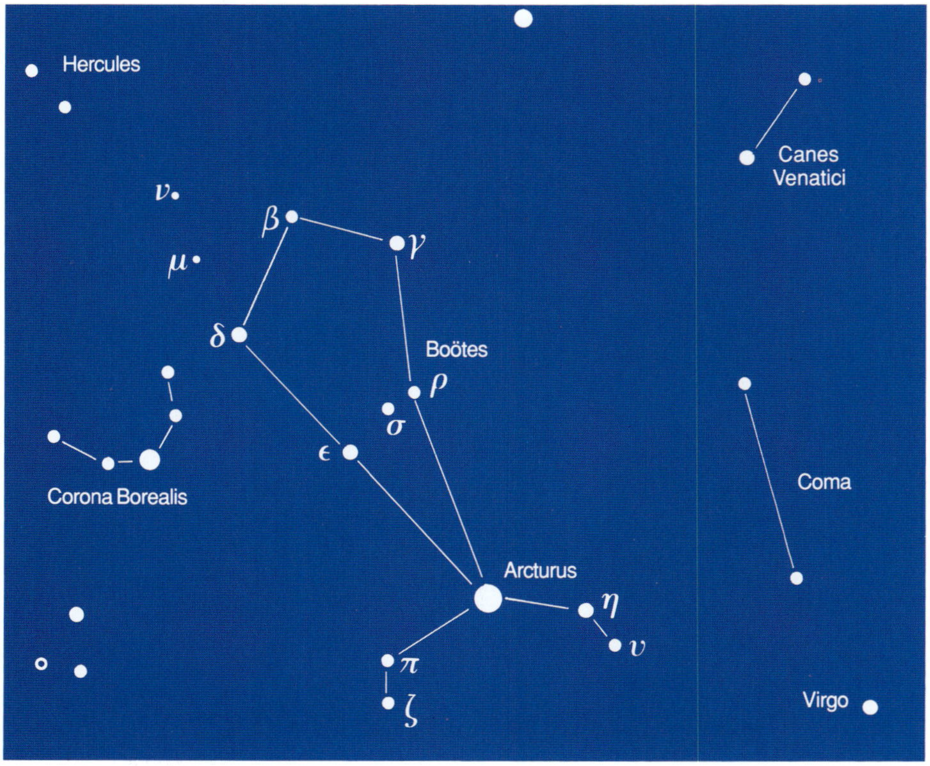

Boötes with Corona Borealis (the Northern Crown).

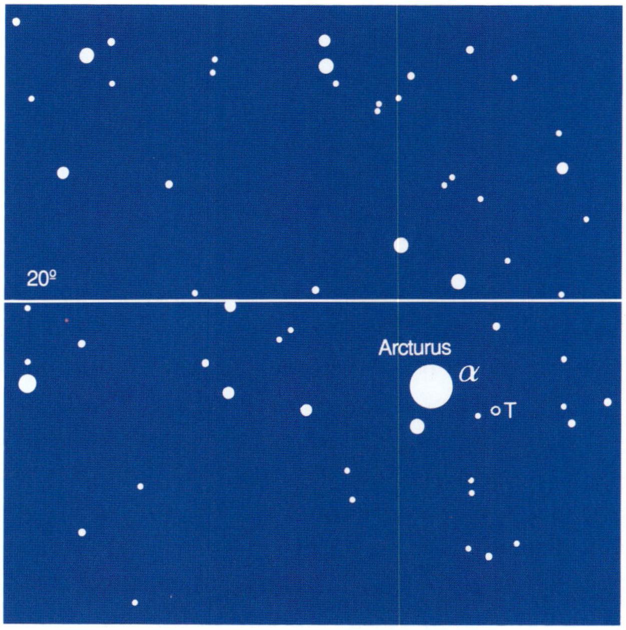

T. Boötis field, showing the bright star Arcturus.

4. ARCTURUS

Alpha Boötis: 16 Boötis

Right Ascension:	14h 15m 39s.6
Declination:	+19° 10' 57"
Apparent magnitude:	-0.04
Absolute magnitude:	-0.2
Spectral type:	K2
Luminosity, Sun=1:	115
Distance, light-years:	36
Parallax:	0".093
Radial velocity:	-5 km/s
Proper motion (per year):	RA -0".08, dec. -2".0

Arcturus is usually said to be the fourth brightest star in the entire sky, but in fact there is a case for placing it third, since it is slightly more brilliant than either individual component of the Alpha Centauri pair. In any case, it is the brightest star in the northern hemisphere of the sky, and its magnitude is slightly above zero.

Since it lies well north of the celestial equator, it is visible from every permanently-inhabited country: only from latitudes south of 76°S is it lost, so that the only living creatures normally denied the sight of it are Antarctic penguins (who probably care very little). It is not circumpolar from the British Isles, but it is on view for part of the night during many months in each year; it is at its best in the spring.

There can be no problem in identifying it. All that has to be done is to follow round the tail of the Great Bear, Ursa Major; this will lead you straight to Arcturus, a lovely orange star whose colour singles it out at once. It is the leader of the constellation of Boötes, the Herdsman, and it dominates the whole area. Its spectrum is of type K.

There are only vague mythological legends associated with Boötes; an old name for it was Arctophylax. It is said that the herdsman was honoured with a place in the heavens for his achievement in inventing the plough drawn by two oxen, but there is nothing definite about it, and Boötes is not linked with any particular character. The Roman writer Vitruvius alluded to it as Custos Arcti, the Bear-Keeper. The name of Arcturus itself is derived from αρκτου ουρα, meaning "Bear's Tail", though elsewhere it is generally referred to as the Guardian of the Bear.

Though Arcturus is a beautiful star, seamen of ancient times regarded it as unlucky, and Pliny even calls it "horridum sidus". Hippocrates, around 460 BC, made much of its supposed effects on the human body, and wrote that a dry season after Arcturus' first appearance in the dawn sky "agrees best with those who are naturally phlegmatic, with those who are of a humid temperament, and with women; but it is inimical to the bilious . . . diseases are especially apt to prove critical in those days." It was referred to by Hesiod, and in the Bible; in the Book of Job we read "Canst thou guide Arcturus with his sons?" In India it was occasionally called Nishtya, or the Outcast, presumably because it lies a long way from the Zodiac. It was also featured in old Chinese star maps, where the sky is divided into asterisms, or groups of stars; each asterism has a determinative star, which need not necessarily be the brightest. Arcturus was one such star, known as Dajido; another was the much fainter Beta Boötis, which we call Nekkar but which the Chinese knew as Qigong.

Arcturus is so bright that keen-sighted people can see it even when the Sun is above the horizon; the German astronomer Julius Schmidt, who spent most of his career in Greece as director of the Athens Observatory and was particularly known for his large map of the Moon, said that he could see it at least twenty-four minutes before sunset. It also seems to have been the first star to have been observed telescopically in broad daylight — by M. Morin in 1635.

Ptolemy called it "golden red", which is reasonable enough. Around 1750 Christian Mayer, who was professor of astronomy first at Heidelberg and then at Mannheim, believed it to be a cluster rather than a single star, which may well indicate that his telescopes were of poor quality; however, in 1777 he did publish a list of double stars which he believed to be binaries, or physically-connected systems. Few people agreed with him at the time, and the existence of binary pairs was not proved until William Herschel's work in 1802.

There is one curious episode which has caused some comment. In 1852 Schmidt reported that Arcturus had lost its usual tinge, and did not return to normal for some years; similar comments were made by two other famous astronomers, Friedrich Argelander and Frederik Kaiser. All three were expert observers; Kaiser was noted for his observations of Mars, while Argelander was largely responsible for a very important star catalogue. Yet it was difficult to

Old figure of Boötes, the Herdsman, with the two Hunting Dogs (Asterion and Chara). Coma Berenices (Berenice's Hair) is to the lower right, and above Boötes is the now-rejected constellation of Quadrans, the Quadrant.

believe in any temporary change in Arcturus; it is simply not that kind of star, and it is very probable that the apparent loss of colour was due to conditions in our own atmosphere.

Arcturus ranks as a giant star, since it is well over 100 times as powerful as the Sun; its diameter is probably about 20,000,000 miles (32,000,000 km), so that if its centre lay in the same position as the centre of the Sun it would more than half-fill the orbit of Mercury. Yet vast though it is, it is still a midget compared with red supergiants such as Betelgeux and Antares; it looks brighter only because it is much closer to us.

Arcturus is a solitary traveller in space; it has no binary companion, and William Herschel suggested that it was isolated enough to be a possible planetary centre. Frankly, this seems most unlikely; if we are to find planets we must look for stars which are not too unlike the Sun, and Arcturus certainly does not qualify. It is at least four times as massive as the Sun, and in the very remote future it may well explode as a supernova, though no catastrophe of this kind is at all imminent!

It has a large proper motion, amounting to 2.28 seconds of arc per year. In 1718 Edmond Halley realized that it had shifted appreciably against the background of more distant stars since ancient times;

Bright Stars of Type K

K-type stars are very common in the Galaxy. Though much less obviously coloured than those of "later" type M, R, N, S, most of them are perceptibly orange, and binoculars bring out the colours well. Here is a list of K-type stars above the third magnitude:

Star	Magnitude	Proper name
Gamma Andromedae	2.14	Almaak
Gamma Aquilae	2.72	Tarazed
Beta Arae	2.85	
Alpha Arietis	2.00	Hamal
Iota Aurigae	2.89	Hassaleh
Epsilon Boötis	2.37	Izar
Epsilon Carinae	1.86	Avior
Alpha Cassiopeiae	2.23 (slightly variable?)	Shedir
Alpha² Centauri	1.39	
Theta Centauri	2.06	Haratan
Beta Cetii	2.04	Diphda
Epsilon Corvi	3.00	
Beta Cygni	3.08	Albireo
Gamma Draconis	2.23	Eltamin
Beta Geminorum	1.14	Pollux
Alpha Hydrae	1.98	Alphard
Gamma Leonis	1.99	Algieba
Beta Ophiuchi	2.77	Cheleb
Epsilon Pegasi	2.38	Enif
Alpha Phoenicis	2.39	Ankaa
Pi Puppis	2.70	
Delta Sagittarii	2.70	Kaus Meridionalis
Lambda Sagittarii	2.81	Kaus Borealis
Epsilon Scorpii	2.29	Wei
Alpha Serpentis	2.65	Unukalhai
Alpha Trianguli Australe	1.92	Atria
Alpha Tucanae	2.86	
Alpha Ursae Majoris	1.77	Dubhe
Beta Ursae Minoris	2.08	Kocab
Lambda Velorum	2.21	Al Suhail al Wazn

The luminosities of these stars have a wide range. Beta Arae is 5000 times as luminous as the Sun, while Haratan could match a mere 17 Suns.

since Ptolemy wrote his *Almagest*, around AD 150, it has crawled across the sky by a distance more than twice the apparent diameter of the full moon (Halley found that Sirius and Aldebaran had also shown detectable shifts). Because of its large proper motion, it was once believed to be the closest of the brilliant stars. This is not so — it is more than four times as far away as Sirius — but by cosmical standards it certainly rates as a near neighbour.

Arcturus is moving through space at about 56 miles per second (90 km per second), in the direction of the constellation Virgo. It

Migrating Stars

The large proper motion of Arcturus will eventually carry it across the boundary from Boötes into Virgo, though not for thousands of years yet. However, there has been one recent case of a migrating star. In 1992 Rho Aquilae, of the fifth magnitude, crossed the border of the adjacent constellation, Delphinus (the Dolphin). So far as naked-eye stars are concerned, it seems that the next wanderer will be Gamma Caeli, in the little constellation of the Graving Tool, which will transfer to Columbus (the Dove) in the year 2400.

Meantime, there is a favourite question to be set in astronomical quiz competitions. "In which constellation is Rho Aquilae?" The answer — "Delphinus!"

must have first become visible with the naked eye half a million years ago, and has brightened steadily since because it is approaching us at 3 miles per second (5 km per second). It is now just about at its nearest, and in the future it will start to draw away, fading until it drops below naked-eye visibility in half a million years' time.

Stellar motions are measured by means of the spectroscope. A star which is approaching us will have the dark lines in its spectrum shifted toward the blue or short-wave end of the rainbow band, while a receding star will show a red shift; this is of course the well-

Herschel's Planetary Centres

It was Sir William Herschel who suggested that Arcturus might well be a planetary centre, because it is a long way from other stars and out of the range of their gravitational attraction. His other candidates, for the same reason, were Vega, Capella, Sirius, Canopus, Markab (Alpha Pegasi), Bellatrix (Gamma Orionis), Menkar (Alpha Ceti), Shedir (Alpha Cassiopeiae), Algorel (Delta Corvi) and Propus (1 Geminorum). In fact, his reasoning was wrong; these stars are not unusually isolated, and Sirius, for example, has a white dwarf companion, while a powerful supergiant such as Canopus appears to be eminently unsuitable as a planetary centre.

William Herschel.

William Huggins and Stellar Radial Velocities

William Huggins was born in London in 1824. At an early age his acquired his first telescope and was fascinated by astronomy, but his family persuaded him to take charge of their drapery business in the City. He devoted himself to this trade between 1842 and 1851, making observations only in his spare time, but eventually he was able to sell the business and move to the suburbs of London, setting up a private observatory at Tulse Hill. From that time onward astronomy occupied his whole life.

At first he concentrated on the planets, using his 8-inch (20-cm) telescope, but he then turned to spectroscopy, working with his friend W.A. Miller, professor of chemistry at King's College. They classified many stars according to their spectra, and were the first to show that the so-called "nebulae" were of two types; those which are made up of stars (now known to be galaxies) and those which are genuinely gaseous. He examined the spectra of comets, and proved that elements such as hydrogen, calcium, sodium and iron were to be found in the stars; he measured many radial velocities, and made the first ultra-violet spectrograph.

Huggins was elected a Fellow of the Royal Society in 1865, and is one of the few amateurs to have served as its President (1900 to 1905). In every way he was a pioneer, and ranks as one of the greatest astronomers of a century ago. He died in 1910.

known Doppler effect. The towards-or-away motion is known as the radial velocity — positive if the star is receding, negative if it is approaching.

The first astronomer who made serious attempts to measure stellar radial velocities was an amateur, William (later Sir William)

T. Boötis

Some old maps still show a star labelled T. Boötis, in the same telescopic field as Arcturus. However, if you try to observe it, you will certainly be doomed to disappointment.

It was recorded on 9 April 1860 by a well-known amateur astronomer, Joseph Baxendell, who was using an excellent 13-inch refractor owned by himself and his friend Robert Worthington. Baxendell's description of it was quite definite. He wrote that "it precedes Arcturus by 1 minute 45 seconds, and is 11 minutes 30 seconds more south. It was mapped down as 9¾-magnitude star on the 9th of April, and on the 11th it was again seen and estimated to be of the 10th magnitude; but on the 22nd it had diminished to 12.8 magnitude, and on the following night it was invisible with Mr. Worthington's 13-inch refractor, in a sky which permitted stars of the 14th magnitude to be seen. It has been frequently looked for since, but not seen."

In fact, it has never been seen again! Energetic searches for it have been carried on, but with a total lack of success. Of course Baxendell could have been mistaken, but he was so experienced an observer that this is hardly likely. It is much more probable that T Bootis, as it has been called, was a fast nova which Baxendell happened to catch at its maximum. Alternatively it could be a recurrent nova, in which case it may reappear in the future; allowing for the proper motion of Arcturus, it should now be 1m 50s west and 7' 50" south of Arcturus. This is really as much as can be said, and T. Bootis remains a mystery.

Stellar Populations

These denote two main types of star regions. Population I consists of brilliant, hot, white or bluish stars, together with inter-stellar material in the form of dust and gas; the brightest stars of Population II are red giants or supergiants, and there is relatively little interstellar material. Since red giants are highly-evolved stars which have used up their main nuclear resources, it follows that Population II objects are older than those of Population I.

No hard and fast boundaries can be laid down, but the arms of spiral galaxies are mainly of Population I, while the central parts of spirals, as well as globular clusters and elliptical galaxies, are mainly of Population II.

Huggins, who set up his observatory in Outer London and carried out pioneer work with his modest 8-inch refractor. In general his results were good, though he was wrong about Arcturus; he believed it to be receding at around 18 km per second rather than approaching us at a much gentler pace.

Arcturus appears to be fast-moving because it belongs to what we call star Population II, made up of a huge spherical system lying round the main flattened Galaxy. It is travelling in a highly inclined orbit, and is cutting through the galactic plane, which is what gives it its high velocity relative to the Sun. Were there any inhabited planets moving round Arcturus, astronomers there would claim that it was the Sun which moves at breakneck speed!

The light of Arcturus was used in a decidedly novel way in the spring of 1933. The "Century of Progress" Exposition was due to open at Chicago, in the United States, and the floodlights were switched on when the light of Arcturus was focused by telescopes on to a photoelectric cell, generating enough electrical current to turn the lights on. Arcturus was selected because at that time it was thought to be 40 light-years away, and the last major exposition in Chicago had taken place exactly forty years earlier.

In 1869 astronomers at the Royal Greenwich Observatory made an attempt to measure the tiny amount of heat reaching us from Arcturus. Using sensitive thermocouples attached to a 12¾-inch refracting telescope, they decided that the heat received was about the same as that which would be received from a 3-inch cube of boiling water from a range of 100 yards. This seems to have been an over-estimate; the real amount of Arcturian heat is about equal to that received from a single candle five miles away. This is not much, but it does stress that Arcturus is a really powerful star, far outmatching our puny Sun.

Radial Velocity

Radial velocity is the toward-or-away motion of a star, and when combined with the star's transverse velocity — i.e. its apparent shift across the sky — it can provide the real speed and direction of movement in space.

If a star is approaching us at 30 miles (50 km) per second, the D lines of sodium at 6000 Ångströms will be shifted to the blue or short-wave end of the spectrum by about 1 Ångström. The Ångström unit, named in honour of the last-century Swedish physicist Anders Ångström, is equal to one hundred-millionth of part of a centimetre.

Paul Doherty

An artist's impression of the Vega dust cloud,

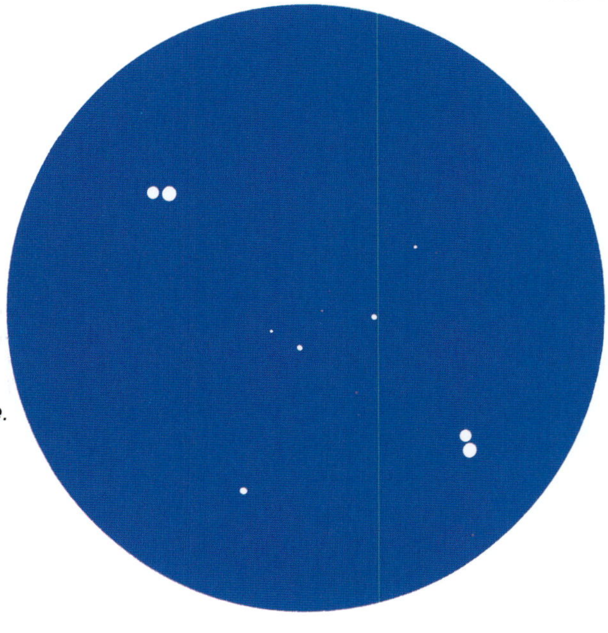

Epsilon Lyrae as seen through a small telescope.

5. VEGA

Lyra.

Alpha Lyrae: 3 Lyrae

Right ascension:	18h 36m 56s.2
Declination:	+38° 47' 01"
Apparent magnitude:	0.03
Absolute magnitude:	+0.5
Spectral type:	A0
Luminosity, Sun=1:	52
Distance:	25 light-years
Parallax:	0".124
Radial velocity:	-14 km/s
Proper motion (per year):	RA +0".017, dec. +0".28

Vega, the second brightest star in the northern hemisphere of the sky — it marginally outshines Capella — is almost overhead as seen from Britain during summer evenings; it is actually circumpolar from London, though at its lowest it brushes the horizon. It is visible at some time or another from anywhere north of latitude 51°S, which means that it can be seen from the whole of New Zealand but is almost out of view from the Falklands. Its great brilliance, together with its steely-blue colour, makes it easy to identify; with Deneb in Cygnus and Altair in Aquila it makes up the unofficial "Summer Triangle".

Vega is the leader of the small but rich constellation of Lyra, the Lyre or Harp, supposed to represent the lyre which Apollo gave to the great musician Orpheus. Aratus called Vega the Little Tortoise, presumably because some lyres were made from tortoise shells; Pliny referred to it as the Harp Star, while in Greece it was sometimes called Cithaea (Χιτηαεα), and to the Arabs it was Nablon, the Phoenician harp. Elsewhere it had avian associations; the Egyptians knew it as the Vulture Star. The name Vega — originally Wega — seems to be derived from the Arabic Al Nasr al Waki, the Swooping Eagle, while Al-Sûfi called it Al Iwazz, the Goose.

Not many star legends have come down to us from Ancient China,

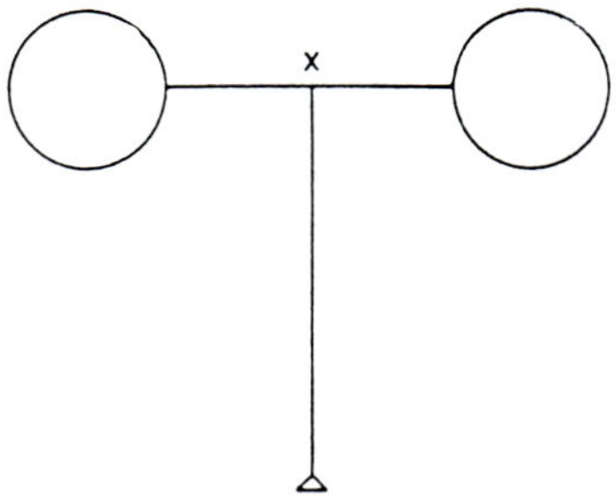

Centre of Gravity. If the two bells of a dumbell are of equal mass, the centre of gravity X will be midway between them. If one mass is greater than the other, the centre of gravity will be displaced toward the more massive bell. This is also the case with the components of a binary system.

but Vega is concerned in one, described in the *Book of Songs* (*Shih Ching*), an anthology of Chinese poetry of the sixth century BC. In this legend Vega, the weaving girl, fell in love with the herd-boy Altair; they were so preoccupied with each other that they neglected their duties, and were punished by being put on opposite sides of the Milky Way. However, they were allowed to meet once a year, on the seventh night of the seventh moon, when a bridge of birds temporarily spanned the "River of Stars" and the lovers could meet.

The Romans used Vega to check their calendar; when it set during the early hours of the morning, autumn was on the way. But much earlier, some 12,000 years ago, Vega was the north pole star, and it will again have this honour in 12,000 years' time. This is because of the effects of precession.

Opposite *The outline of Lyra is clear, with Vega outstanding.*

Below *The Great Pyramid. Certainly the Pyramids are astronomically aligned; the position of the celestial pole has shifted since they were built.*

The Earth is not a perfect sphere; its equatorial diameter is 26 miles (42 km) greater than the diameter as measured through the poles, so that in effect the equator bulges out. The Sun and Moon pull on this bulge, and the result is to make the Earth wobble, or "precess", rather in the manner of a boy's gyroscope which is running down and has started to topple. The direction of the axis changes, and this affects the positions of the celestial poles; each moves in a circle in the sky, taking 26,000 years to complete one

Photograph: Patrick Moore.

Photograph: R.W. Arbour

The Lyrid Meteors

The Lyrid meteor shower, active between 19 and 25 April each year, radiates from a point close to Vega. Normally it is sparse, but sometimes, as in 1922, it can be very rich.

Meteors are cometary debris. As a comet moves along, it leaves a dusty trail behind it; if the Earth passes through one of these trials, it collects a great deal of debris. The tiny particles, usually smaller than grains of sand, dash into the upper air, travelling at anything up to 45 miles (72 km) per second; they rub against the air-particles, and are so violently heated that they burn away in the streaks of luminosity which we call shooting stars. Of course, what we are actually seeing is not the tiny particle itself, but the effects which it produces during its headlong plunge through the air. Meteors burn out before they penetrate to below 40 miles (65 km) above sea-level, and end their journey to the ground in the form of ultra-fine dust. Larger bodies may survive the complete drop and land intact, sometimes producing craters; these objects are termed meteorites — but they come from the asteroid belt, and are not associated with comets or with shooting-stars.

The meteors of any particular shower seem to radiate from a definite point in the sky, because they are moving through space in parallel paths; it is an effect of perspective. To show what is meant, stand on a bridge overlooking a motorway and see how the parallel lanes seem to radiate from a point near the horizon. Showers are named according to the constellation which contains the radiant; the August Perseids from Perseus, the December Geminids from Gemini, and so on. The April meteors come from Lyra. Among the main annual showers are:

Name of shower	**Begins**	**Maximum**	**Ends**	
Quadrantids	1 Jan	4 Jan	6 Jan	Radiant in Boötes, Short, sharp maximum. Occasionally rich
Lyrids	19 Apr	21 Apr	25 Apr	Sometimes rich.
Eta Aquarids	24 Apr	5 May	20 May	Broad maximum.
Perseids	23 July	12 Aug	20 Aug	Richest annual shower.
Orionids	16 Oct	22 Oct	27 Oct	Swift meteors.
Leonids	15 Nov	17 Nov	20 Nov	Usually sparse, but occasional 'storms' (1599, 1833, 1866, 1966)
Geminids	7 Dec	13 Dec	16 Dec	Generally rich.

The Quadrantids are so named because the radiant lies in a area once making up a constellation called Quadrans (the Quadrant), which has been dropped from modern maps.

Most showers are associated with known comets; the Eta Aquarids and the Orionids with Halley's Comet, the Perseids with Comet Swift-Tuttle, the Leonids with Comet Tempel-Tuttle, and so on. The April Lyrids are linked with Thatcher's Comet, which was seen in 1861 and reached naked-eye visibility, but will not return for 415 years or thereabouts. The comet was discovered by the American amateur astronomer A.E. Thatcher , and has no connection with any modern politician!

revolution. When the Egyptian Pyramids were being built, the north polar star was Thuban in the constellation of the Dragon; today it is Polaris in the Little Bear. Vega's spell as pole star ended in prehistoric times, but to our remote descendants — if mankind still inhabits the Earth — Vega will regain its proud position.

Though Lyra is a small constellation, it contains a surprising number of interesting objects. Close by Vega is Epsilon Lyrae, which keen-eyed people can see to be made up of two; with even a modest telescope each component can again be split, so that Epsilon Lyrae is a double-double or quadruple star. Zeta Lyrae, also close to Vega, is an easy telescopic pair. Beta Lyrae, or Sheliak, is an eclipsing binary ranging between magnitudes 3.3 and 4.3, and is known to be a very complex interacting system. Between it and the adjacent third-magnitude Gamma Lyrae is Messier 57, the most famous example of a planetary nebula — inappropriately named, because a planetary nebula is not truly a nebula and has absolutely nothing to do with a planet; it is an old star which has thrown off its outer layers, and is surrounded by a shell of expanding gas. Telescopically, M.57 looks like a tiny luminous cycle-tyre, with the remnant of the old star in its centre. But the whole region is dominated by the steely blue brilliance of Vega.

At a mere 26 light-years it is one of the nearest of the first-magnitude stars, and it was also one of the first to have its distance measured; F.G.W. Struve, in Russia, measured its parallax at about the same time that Bessel was studying 61 Cygni, and although Struve was less accurate — because Vega is further away — his result was at least of the right order. It is worth noting that William Herschel had looked at Vega through his telescopes and believed that it showed an apparent diameter of one-third of a second of arc, but in this he was wrong; in any ordinary telescope a star appears as virtually a point source of light.

Vega has two faint companions, one of the tenth magnitude and the other of the twelfth, but they are not genuinely associated with Vega or with each other; they are much further away, and simply happen to lie in much the same direction as seen from Earth. Incidentally, the Sun is moving through space toward a point not far from Vega; this point is termed "the Apex of the Sun's Way".

Vega is a typical Main Sequence star of spectral type A. It is much hotter than the Sun, with a surface temperature of over 9000°C, and is three times as massive; its diameter has been estimated as 2,300,000 miles (3,700,000 km), and as well as being more than fifty times as luminous as the Sun it is also more energetic. It is of special interest for another reason too; it has been found to be associated with cool material which may even be planet-forming. The discovery was made in 1983 with IRAS, the Infra-Red Astronomical Satellite, one of the most successful of near-recent space missions.

IRAS was launched by rocket in the early hours of 26 January 1983. Before long it had settled in its orbit, moving round the Earth at a height of approximately 560 miles (900 km) in a period of 103 minutes. It was put into a path which carried it over both the poles, and therefore stayed close to the demarcation line between the day and night hemispheres. The advantage of this so-called sun-synchronous orbit is that the satellite is in sunlight for most of the time, and can draw power from its solar panels.

The information from IRAS was collected and stored on a tape. It was then played back each time the satellite passed over the re-

F.G.W. Struve

Friedrich Georg Wilhelm Struve was the first of a whole line of eminent astronomers; indeed, the Struve family is perhaps the most famous in astronomical history apart from the Herschels. F.G.W. Struve was a German, born in Altona, near Hamburg, in 1793. He went to Estonia, then part of the Russian Empire, and studied at Dorpat University; in 1818 he was appointed Director. He obtained an excellent 9-inch refracting telescope made by Fraunhofer; it was clock-driven (a rarity in those days) and Struve used it to great effect, discovering over 2000 new double stars. It was while at Dorpat that he measured the parallax of Vega; his work was quite independent of Bessel's or Henderson's. In 1839 he went to Pulkova Observatory as Director, and continued his double star work; he also took a great interest in cosmology and in the absorption of light in interstellar space. He retired in 1861, and died three years later; he was succeeded as Director at Pulkova by his son Otto.

Optical Double Stars

Double stars are of two kinds: binaries and optical pairs. Binaries are physically-associated systems, with components moving round their common centre of gravity, and which were presumably born from the same mass of material. With an optical double, there is no real connection; one of the components is in the background, so to speak, and the apparent proximity is due to a line of sight effect. Rather surprisingly, binaries are much commoner than optical pairs.

IRAS, the Infra-Red Astronomical Satellite, which operated for much of 1993 and detected the infra-red excess round Vega.

ceiver at the Rutherford Appleton Laboratory at Chilton, in Oxfordshire (England). The main instrument was a 22.4-inch infra-red telescope, which looked very much like an ordinary optical telescope, and collected its radiation by means of a curved mirror. Of course it could not produce a visible picture, since infra-red radiation does not affect our eyes, and there was another problem also; by infra-red standards, a telescope at room temperature is very "hot", so that its own radiation will completely mask the much weaker signals coming from the sky. Therefore the instrument has to be cooled, and this is done by using liquid helium, which boils at a temperature of -270°C, only just above absolute zero (that is to say, the lowest temperature there can possibly be). Unfortunately the liquid helium boils steadily away, and this was why IRAS operated for less than a year. However, it accomplished a great deal, and it was amazingly sensitive. It would have been capable of detecting the radiation from a warm football over a distance equal to that between London and New York.

Two of the research scientists at Chilton, Drs. Hartmut Aumann and Fred Gillett, were calibrating the IRAS telescope, and using Vega as a source, when they found that Vega had what was called "a huge infra-red excess"; it was emitting much more strongly than had been expected. Evidently it was surrounded by a cloud of cool material, and further studies showed that this cloud was made up of particles which were considerably larger than the tiny dust particles found in interstellar space. The cloud extended out to 7,400,000 miles (12,000,000 km) from the star, which is about 80 times the distance between the Earth and the Sun. Aumann commented; "If there were small particles round Vega, there must be large particles also. Very careful measures have been made of what are called fragmentation products. When I applied these methods in the case of Vega, I was astonished, because it seemed that the total mass was much the same as that of all the planets in our Solar System combined." The temperature of the cloud particles was given as -184°C, about the same as the temperature of the icy particles in the rings of Saturn.

Drs. Aumann and Gillett, at Chilton in Oxfordshire, where they detected the Vega infrared excess from the IRAS.

Deceptive Pairs of Stars

Pairs of stars are not always what they seem. Vega's two optical companions are faint, but there are much more striking examples of unconnected components. For example Algiedi or Alpha Capricorni, in the Zodiacal constellation of the Sea-goat, is made up of two components, of magnitudes 3.6 and 4.2 respectively; they can be seen separately with the naked eye. The fainter star is 1600 light-years away and is very luminous; the brighter member lies at a mere 117 light-years.

Delta Gruis, in the southern constellation of the Crane, is another example; both the components are of the fourth magnitude, but their distances are respectively 155 and only 88 light-years. Zeta Scorpii, in the Scorpion, is made up of two stars, magnitudes 4.7 and 3.6, at distances of 2500 and only 163 light-years. But perhaps the most striking pair is again in the Scorpion, made up of Lambda Scorpii or Shaula (magnitude 1.6) and Upsilon Scorpii or Lesath (2.7). Shaula is 275 light-years from us, and 1300 times as powerful as the Sun — but Lesath is 1560 light-years away, and could match 16,000 Suns. This means that Lesath is a great deal further away from Shaula than we are. There are, however, some wide pairs which are physically associated, such as Theta and Delta Tauri in the cluster of the Hyades.

The most famous naked-eye pair is, of course, in the Great Bear; Mizar has a fourth-magnitude companion, Alcor. The two are a long way apart, but they share a common motion through space, and presumably have a common origin.

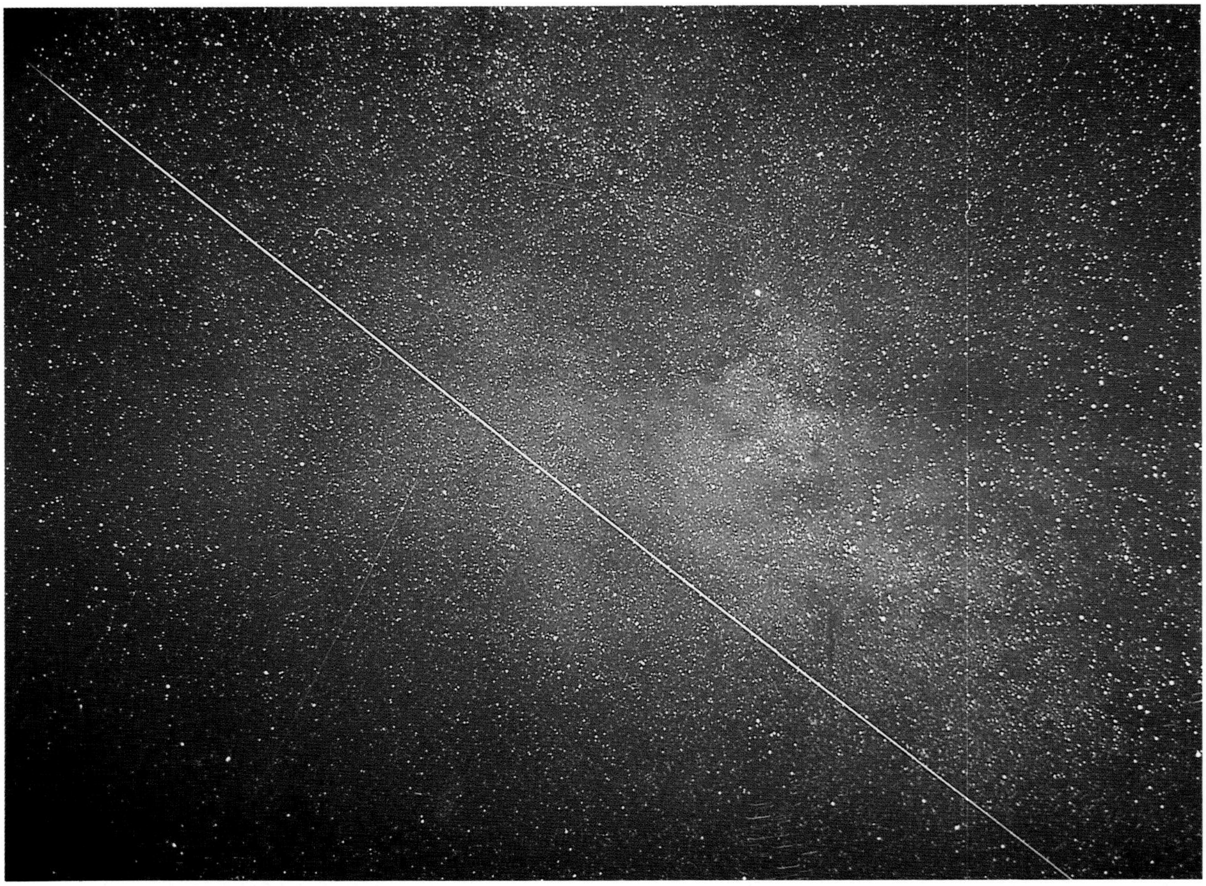

Brilliant meteor, 12 August 1993, 0259 GMT.
Photograph: Michael Maunder

So does Vega have a system of planets? It would be premature to claim anything of the kind, but it does seem definite that potentially planet-forming material exists. Yet remember that Vega is much hotter and more active than the Sun, and sends out much more ultraviolet radiation, which would be harmful to living creatures of our type. Even if we could prove that there are Vegan planets, we would still have no clue as to whether there could be any Vegan astronomers!

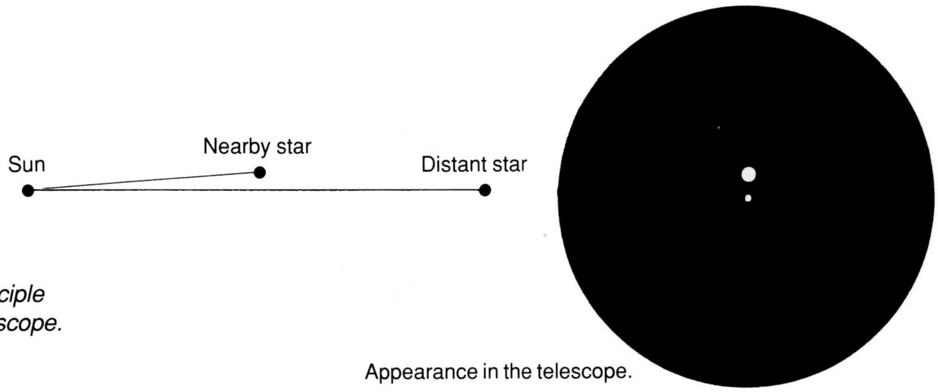

The optical double principle as seen through a telescope.

6. CAPELLA

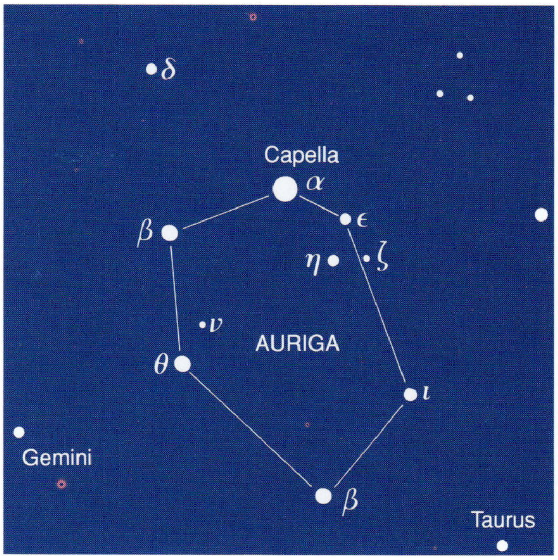

Auriga.

Alpha Aurigae : 13 Aurigae

Right ascension:	05h 16m 41s.3
Declination:	+45° 51' 53"
Apparent magnitude:	0.08
Absolute magnitude:	0.3
Spectral type:	G8+ FO
Luminosity, Sun=1:	90+70
Distance, light-years:	43
Parallax:	0".073
Radial velocity:	+30 km/sec
Proper motion (per year):	RA +0".008, dec. -0".42

Capella is the northernmost of the first-magnitude stars; even so, it is visible at one time or another from almost all inhabited countries, and is permanently lost only from the southernmost tip of New Zealand and outposts such as the Falkland Islands. In brightness it is virtually equal to Vega, but it is of very different colour; its yellowish tinge contrasts sharply with the steely blue of Vega. Both are circumpolar over the British Isles; they lie on opposite sides of the north celestial pole, and at roughly the same distance from it. This means that when Capella is high up, Vega is low down, and vice versa. From Britain, Capella is almost overhead during winter evenings, while in summer this proud position is occupied by Vega.

Capella is the leader of the large, prominent constellation of Auriga, the Charioteer, which covers over 650 square degrees of the sky and contains a variety of interesting objects. In mythology it is said to represent Erechthonius, son of Vulcan, the blacksmith of the gods, who invented the four-horse chariot and subsequently became King of Athens. Manilius, a Roman poet who died about 50 BC, wrote:

> Near the bent Bull a Seat the Driver claims
> Whose skill conferr'd his Honour and his Names.

Old figure of Auriga, the Charioteer.

> His Art great Jove admir'd, when first he drove
> His rattling Car, and fix'd the Youth above.

The bent Bull is Taurus, which lies beside Auriga and has annexed one of Auriga's stars! Capella itself is known as the She-goat, and the three faint stars which form a triangle close beside it are the Haedi, or Kids. In Chinese maps it was one of five stars making up Woo Chay, the Five Chariots (the others were Beta, Theta, Kappa and Gamma Aurigae); to the Arabs it was Al Rakib, the Driver, because in its far-northern position in the sky it was often visible in the early evening before any other stars came into view. Astrologically, it was linked with wealth and with military honours.

It is worth noting that a few of the old observers referred to Capella as reddish. Ptolemy was among these, and so, much later, was Riccioli, an Italian Jesuit who is best remembered for his map of the Moon, completed in 1651. However, there is not the slightest chance that there has been any colour-change in historic times. Undoubtedly the observations were made with Capella low over the horizon, so that the off-white hue was accentuated.

Almost a century ago now, in 1899, an interesting discovery was made. The spectrum of Capella was examined, by W.W. Campbell at the Lick Observatory in America and by H.F. Newall at Cambridge in England, and was found that Capella was not a single star; it is a binary, but the components are so close together that they are excessively hard to separate visually. Campbell and Newall made their discovery by using the spectroscope. Changes in the appearance of the dark absorption lines, due to the Doppler effect, revealed

The Kids

The three stars forming the triangle of the Haedi or Kids may look obscure, but two of them are of special interest. The third, Eta Aurigae, is normal enough, but Epsilon and Zeta are eclipsing binaries, each unique in their own ways.

Epsilon, the more extreme, has no generally-used proper name, though it has been called Almaaz. It is a particularly luminous supergiant, as powerful as Canopus, and usually it shines as a star of the third magnitude. As long ago as 1821 a German astronomer named Fritsch found it to be variable, and much later it was classified as an eclipsing binary of remarkably long period — 9892 days, or about 27 years. At the onset of an eclipse there is a long, slow fade to magnitude 2.8, followed by a prolonged minimum and then an equally slow recovery. The last eclipse began on 22 July 1982; it was total from 11 January 1983 to 16 January 1984, and ended on 24 June 1984, so that nothing more will happen until the year 2011.

The strangest point is that the eclipsing secondary has never been seen. It shows no spectrum, emits no infra-red radiation and is silent at radio wavelengths. But for the eclipses, we would have no clue about its existence. So what precisely is it? It was once thought to be a huge "proto-star", not yet hot enough to shine, and big enough to fill the Solar System out almost to the orbit of Uranus, in which case it would have been the largest star known, but it is now thought more likely to be a much smaller star surrounded by a shell of opaque or semi-opaque material. It is a pity that eclipses are so rare; we may learn more in 2011.

By sheer chance the third Kid, Zeta or Sadatoni, is also an eclipsing binary of long period — in this case 972 days; the range is between magnitudes 3.7 and 4.2, so that the variations are not striking. Here we have a K-type supergiant primary with a smaller, hotter companion. It is when the hot star is hidden (or partially hidden) by the supergiant that we see a drop in light. Spectroscopically these eclipses are fascinating. When they begin or end, there is a period when the light of the hot star comes to us after passing through the rarefied outer layers of the supergiant.

The fact that these two unusual binaries lie side by side in the sky is due to nothing more significant than a line of sight effect. There is absolutely no connection between them; Zeta lies very much in the foreground, so to speak.

Capella's true nature; it is what is termed a spectroscopic binary (I will have more to say about these when discussing Spica, which is of the same class).

Both components are yellow, and in colour not unlike each other; the brighter of the two is the type G, while the fainter is either late G or early F. Both are large, with diameters respectively of 11,200,000 miles (18,000,000 km) and 6,000,000 miles (9,600,000 km). In mass they are very similar; the G-star has 3.09 times the Sun's mass, and the F-star 2.95 Suns (remember that a smaller star is always denser than a larger one). The real distance between the two is 70,000,000 miles (11,300,000 km), and they move round their common centre of gravity in a period of 104 days. The orbits are practically circular.

The distance between the components is less than that between the Earth and the Sun, and since Capella is so far away it is clear that the apparent separation must be very small indeed. As soon as the binary nature was discovered, efforts were made to split the components optically. Partial success came in 1901, when astronomers at Greenwich, using the 28-inch refractor, saw the star as "elongated". Confirmation came much later, with the aid of the 100-inch Hooker reflector at Mount Wilson; the separation ranges between 0.04 and 0.05 of a second of arc. This has been amply re-confirmed

Transferred Stars

Auriga may be large and prominent, but it has lost one of its leading stars. This is Al Nath, of magnitude 1.65; it is hot and white, with a B-type spectrum, and is 400 times as luminous as the Sun. Its distance is 130 light-years. It used to be listed as Gamma Aurigae, and this was local enough, as it clearly belongs to the Auriga pattern. However, in the revision of the constellations undertaken by the International Astronomical Union, it was transferred to the neighbouring Taurus, the Bull, as Beta Tauri — and Beta Tauri it has remained.

(It is not the only case. For example Gamma Scorpii has become Sigma Librae, while Alpheratz, in the Square of Pegasus, has changed from Delta Pegasi into Alpha Andromedae.)

However, Auriga still has many bright stars left. Those above the fourth magnitude are:

Star	Magnitude	Spectrum	Luminosity Sun+1	Distance, lt-yrs
Alpha (Capella)	0.08	G+F	90+70	43
Beta (Menkarilina)	1.90	A2	50	46
Theta	2.62	AO	75	82
Iota (Hassaleh)	2.69	K3	700	267
Epsilon (Almaaz)	2.99v	FO	200,000	4600
Eta	3.17	B3	450	200
Delta	3.72	KO	60	163
Zeta (Sadatoni)	3.75v	K4	700	520

Paul Doherty

An artist's impression of a starship approaching the Epsilon Aurigae system.

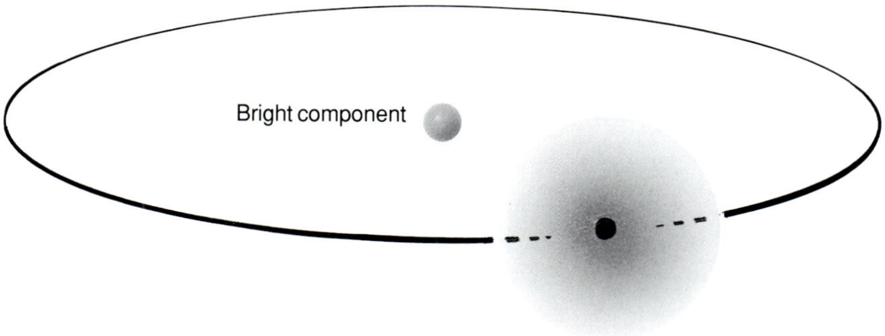

The theory of Epsilon Aurigae.

Eccentricity

It is eccentricity (denoted by e or ε) which determines the shape of an orbit. An ellipse has two foci, and the eccentricity depends on the ratio between the separation of the two foci and the longest "diameter" of the ellipse, known as the semi-major axis. For a circle the foci coincide, so that the eccentricity is zero. For a parabola the eccentricity is 1, so that a parabola is an open curve; for hyperbolic shape the eccentricity exceeds 1.

The eccentricity of the Earth's orbit is 0.17, which means that it is almost circular; for the components of Capella the eccentricity is only 0.01, so that if the orbits were drawn to a scale fitting a page of this book they would be virtually indistinguishable from circles. Comets moving round the Sun often have eccentricities exceeding 0.9, and of course there are also comets moving in open curves, so that after passing through perihelion they never return.

The Mount Wilson 100-inch reflector; for a long time the most powerful in the world.
Photograph Patrick Moore, 1994

with modern electronic devices, but Capella remains an excessively difficult visual double, and normal telescopes show it as a single point of light. It does, in fact, provide something of a link between purely spectroscopic binaries, too close to split optically, and normal pairs.

Some new and very interesting observations of Capella were made in February 1996 from Cambridge Observatory, in England. Use was made of a new type of optical telescope developed there; it is so sensitive that it could make out detail in stars at the same level needed to read a car registration plate at a distance of 1000 yards (914 metres). It is called COAST — short for the Cambridge Optical Aperture Synthesis Telescope.

COAST combines the light beams from an array of small telescopes to give the effect of a telescope equal in size to the overall distance between the array elements. To increase the detail, all that has to be done is to move the array telescopes further apart. The same principle has long been used in radio astronomy, but optical waves are a million times shorter than radio waves, so that much greater stability and precision is needed. With COAST there are three telescopes, 6 yards (5.5 metres) apart (a fourth telescope will soon be added). The current design allows for a separation of 100 yards (91 metres), giving detail in the images down to 1/1000 of a second.

Capella was the first target for Dr. John Baldwin and his team at Cambridge. The components of Capella are a mere 6 light-minutes apart, and the apparent separation is only 1/20 of a second of arc, but for the first time the two Capellas were seen clearly separated.

There is a third member of the system — a faint red dwarf; rather confusingly it is known as Capella H, because the earlier letters of the alphabet had already been used for background stars totally unconnected with Capella. The dwarf is 0.17 of a light-year from the main pair, which amounts to 11,000 times the distance between the Earth and the Sun. It too is double, with each component having only a tiny fraction of the Sun's luminosity. It is none too easy to identify even when its position is known; it lies about 12 minutes of

The Sky from Capella

In 1975 J. Holmes gave a description of the sky from Capella.* Cassiopeia, he wrote, would look rather like a twisted paper-clip. The leading stars, with their distances in light-years, would be:

Star	Apparent magnitude	Distance, lt-yrs
Aldebaran	-0.45	37
Canopus	-0.38	115
Rigel	0.08	875
Castor	0.15	23
Betelgeux	0.26	485
Beta Aurigae	0.34	43
Pollux	0.43	24
Achernar	0.90	141
Algol	1.06	66
Antares	1.09	550

However, there must be uncertainties; for example Holmes was using a distance for Canopus which was much less than that given in the Cambridge catalogue which I am following here.

**Sky and Telescope, 49,* 147 (1975)

arc south-east of Capella at a position angle of 141 degrees, but the combined magnitude is only 10.

It seems most unlikely that there are any planets in the Capella system; if there are, any planets must be a long way from the twin giants. Certainly a Capellan astronomer would have a fascinating view, but it would be very unfamiliar to beings such as ourselves. On the other hand the constellation patterns would in many cases be recognisable to us, because Capella is only 43 light-years away, and although this sounds (and is!) a great distance it is not very far on the scale of the Galaxy. Our Sun would be rather below naked-eye visibility — assuming, of course, that our hypothetical Capellans have sight similar to ours. The brightest stars would be Aldebaran, Canopus, Rigel and Castor.

It is impossible to overlook Capella. It has often been said to be a "solar-type" star, but there is one vitally important difference. Our Sun is a yellow dwarf, while Capella is a yellow giant — or, rather, a pair of yellow giants.

7. RIGEL

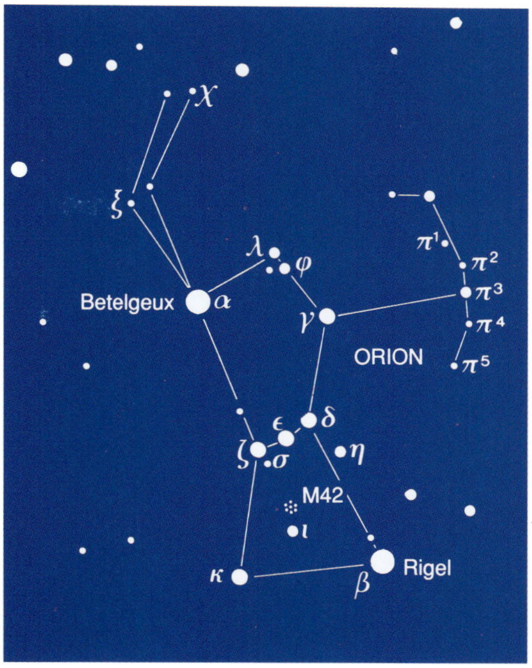

Diagram of Orion, showing Betelgeux and Rigel.

Beta Orionis: 19 Orionis

Right ascension:	05h 14m 32s.2
Declination:	-08° 12' 06"
Apparent magnitude:	0.12
Absolute magnitude:	-7.1
Spectral type:	B8
Luminosity, Sun=1:	60,000
Distance, light-years:	910
Parallax:	0".002
Radial velocity:	+21 km/s
Proper motion (per year):	RA 0".00, dec. 0."00

Rigel is only the seventh brightest star in the sky, but in sheer power it far outstrips most of its rivals. There may be some doubts about the immense luminosity of Canopus, but there are none at all about Rigel, which is at least 60,000 times as powerful as the Sun.

It is the leader of Orion, the Hunter, though it has been given the second Greek letter; normally it is distinctly brighter than Alpha Orionis (Betelgeux). Orion itself is perhaps the most magnificent of all the constellations, and the pattern of seven bright stars is striking by any standards. Since it is crossed by the celestial equator, it is visible from almost everywhere in the world. Actually the equator passes close by Mintaka or Delta Orionis, the northern member of the three stars making up the Hunter's Belt. Only from the extreme polar regions is Orion not visible in its entirety; it is perhaps a little unexpected to find that from the South Pole, where a major observatory is now being set up, Rigel is permanently above the horizon and Betelgeux never.

Many legends are associated with Orion, and one of these, from Greece, is not entirely to his credit. Apparently he was chasing seven beautiful girls, with intentions which were far from honourable, when the gods intervened, changing the girls into stars and swinging them

Opposite *Orion trails. This is a time exposure picture and the redness of Betelgeux contrasts with the whiteness of Rigel. The Orion Nebula also looks red, though this is not evident visually.*
Photograph: Christopher Doherty

up into the sky to become the Pleiades. It is also said that Orion boasted of being able to kill any creature on earth — but he forgot the scorpion, which crawled out of a hole in the ground, stung him in the heel and killed him. When Orion was given a place in the heavens, it seemed only fair to put the scorpion there also — but the two were set as far apart as possible, so that they can never meet.

Orion was of great importance in Egyptian lore; he was Sahu, the soul of Osiris, the ruler of the underworld. In the famous Dendereh Zodiac he is shown journeying through the sky in his celestial boat, followed by Sothis (Sirius), soul of the goddess Isis. To the Arabs, Orion was Al Jauzah, the Giant. The constellation was generally linked with bad weather, and in the second century BC the Roman poet Manilius commented that the Roman fleet was destroyed by storms, during the First Punic War against Carthage, because the commander had made the mistake of putting to sea "at the time of the rising of Orion".

The name "Rigel" is Arabic, and is derived from Rijl Jauzah al Yusra, the Left Leg of the Giant. To Al-Sûfi, who drew up an excellent star catalogue, it was Rai al Jauzah, the Giant's Herdsman, with four camels represented by other stars in the constellation (Betelgeux, Bellatrix, Mintaka and Saiph). Older names were Algebar, Regal and Riglon, all of which have long since fallen into disuse. The modern form was introduced in the Alphonsine Tables, published in 1521.

The Stars of Orion

Apart from Betelgeux, all the leading stars of Orion are hot and white or bluish-white. Those above the third magnitude are:

Star	Magnitude	Spectrum	Luminosity, Sun=1	Distance, lt-yrs
β Rigel	0.12	B8	60,000	910
α Betelgeux	0.5v	M2	15,000	310
γ Bellatrix	1.64	B2	2200	360
ε Alnilam	1.70	O7	23,000	1200
ζ Alnitak	1.77	O9.5	19,000	1100
κ Saiph	2.06	B0	49,000	2100
δ Mintaka	2.23v	O9.5	22,000	2350
ι Hatysa	2.76	O9	20,900	1860

Saiph is therefore not greatly inferior to Rigel. A million years ago its magnitude was -4.3, and it shone as the most brilliant star in the sky.

Closely north of Hatysa is the Great Nebula in Orion, M.42, a colossal mass of dust and gas inside which fresh stars are being born. It is illuminated by the stars of the Trapezium, Theta Orionis, so nicknamed because of the arrangement of its four main components. Even a small telescope will give a glorious view of it.

J.S. Plaskett

John Stanley Plaskett was a Canadian, born in 1865. He trained as an engineer, but then turned to astrophysics, and became Director of the Dominion Observatory at Victoria, British Columbia. He specialized in measurements of stellar radial velocities, and his work led him on to a confirmation of the rotation of the Galaxy around its centre, now known to be almost 30,000 light-years away. Plaskett found many new double stars, and one of these binaries, still known as Plaskett's Star, is particularly massive; the combined mass of the two components is at least 90 times that of the Sun. Plaskett retired in 1935, and died in 1941.

Rigel's spectrum is given as B8. Each spectral type is separated into divisions, from zero to 9; the initial sequence, as we have noted, is, O, B, A . . . so that Rigel is eight-tenths of the way from B0 to A0. It is white, with a surface temperature not much below 10,000 degrees. The internal temperature is much greater than that of the Sun, and may be around 100 million degrees; the energy is being produced by the conversion of helium into carbon, so that most of the original hydrogen "fuel" has been used up. It is over 900 light-years away, so that we now see it as it used to be in the time of William the Conqueror. If it were as close to us as Sirius, it would have an apparent magnitude of -10, and would send us at least one-fifth the light of the full moon.

In 1908 J.S. Plaskett, one of the great pioneers of astrophysics, examined the spectrum of Rigel and found something very curious. There were unexpected variations which led him to suggest that Rigel might be a very close binary — as was already known to be

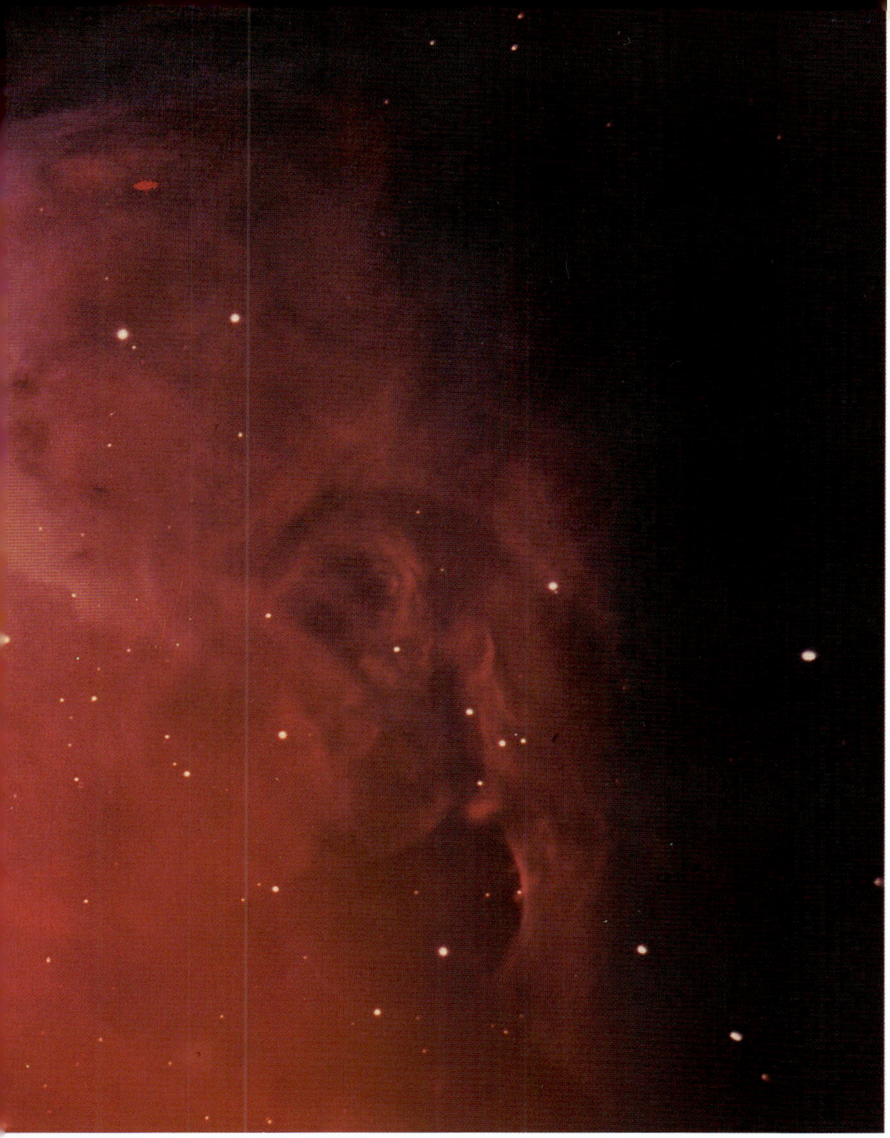

The Orion Nebula.
200-inch photograph, Palomar

true of Capella. Naturally, efforts were made to split the pair visually, and very large telescopes were used, notably the 40-inch refractor at the Yerkes Observatory — then, as now, the largest refractor in the world. There had been success with Capella, but none with Rigel, and it now seems that Plaskett made one of his rare errors of judgement. Rigel is not a spectroscopic binary; it is single, and the spectral variations are due to changes in the star's atmosphere. Remember, Rigel is highly luminous, and much more energetic than our sober, mild Sun.

Yet Rigel is not a solitary traveller; it has a companion, discovered in the 1830s by F.G.W. Struve from Dorpat in Estonia (the same Struve who made a praiseworthy, albeit rather inaccurate, measurement of the parallax of Vega). The companion, known officially as Rigel B, lies 9.5 seconds of arc from the main star, at a position angle of 202 degrees; since its magnitude is 6.8, it would be an easy object if it were not so overpowered by the brilliance of the primary. Even

The Yerkes 40-inch refractor — still the largest in the world.
Photograph: Patrick Moore, 1994

The Yerkes Refractor

The Yerkes Observatory was set up in 1897, at Williams Bay in Wisconsin, USA. It was put there at the request of the millionaire C.T. Yerkes, who had provided most of the money for the main telescope. This is a refractor with an object-glass 40 inches (1.01 metres) in diameter. It remains the largest telescope of its kind; with greater aperture, an object-glass would tend to distort under its own weight, making it useless, so that all the great modern telescopes are of the reflecting type. However, the Yerkes 40-inch remains in regular use.

The Observatory itself adjoins a golf course, and could not be more unlike most other great astronomical centres. Williams Bay is much more accessible than, say, the summit of Mauna Kea in Hawaii, or the wilds of the Atacama Desert in Northern Chile!

so it is by no means difficult, and I have often seen it with a modest 3-inch (7.5-cm) refractor. It is a long way from Rigel — over 2000 times the distance between the Earth and the Sun — and there has been no perceptible change in the separation or position angle since Struve first saw it, but there is little doubt that it and Rigel are genuinely associated, because they share a common motion through space.

In 1937 astronomers at Mount Wilson, using the 100-inch Hooker reflector, found that B is a spectroscopic binary; we now refer to the two components as B and C. Both are quite massive — 2.5 and 1.9 times more powerful than the Sun respectively. They are also luminous; B could match 100 Suns, so that it is nearly the equal of Arcturus, while C has about half this luminosity. Each has a B-type spectrum. They take 9.9 days to move round their common centre of gravity, in almost circular orbits.

There is a minor mystery associated with the pair. Several famous double-star observers of the early twentieth century, notably S.W. Burnham and R.G. Aitken, reported that they could, under excellent conditions, see the components separately. However,

Dome of the Yerkes 40-inch refractor.

spectroscopic work has shown that they are much too close; they can never be as much as a tenth of a second of arc apart, so that the reports must have been wrong, just as Plaskett had been wrong in claiming that Rigel itself was a binary system.

Because Rigel is so powerful, it is capable of lighting up dust-clouds a long way from it — such as the nebula IC 2118 in the constellation of Eridanus, even though the nebula is more than two degrees from Rigel in the sky. Rigel is also a member of a group of hot stars known as the Taurus–Orion Association; very probably the members of this group had a common origin.

Rigel is very massive; it "weighs" as much as at least 25 Suns, and this leads on to an interesting speculation. When a star of this kind comes to the end of its nuclear reserves, it does not simply subside into a white dwarf, as Sirius B has done and the Sun will eventually do. There is an "implosion", followed by an explosion, and the star literally blows itself to pieces in what is termed a supernova outburst. Until recently it was thought that only a red supergiant could "go supernova", but this is now known to be wrong. In 1987 a supernova flared up in the Large Cloud of Magellan, a southern galaxy 169,000 light-years away, and reached naked-eye visibility. The precursor star could be identified, and turned out to be not a red giant, but a blue one. So could the same fate befall Rigel? It is certainly possible, though not for a long time yet. However, massive stars run through their life-stories much more quickly than dwarfs, and Rigel cannot go on shining in its present form for more than ten million years at most. Meanwhile, it is there for our inspection and admiration — one of the real searchlights of the Galaxy.

The Alphonsine Tables

The Alphonsine Tables, in which the name of Rigel was first used in its current form, were compiled at the command of King Alphonso X of Castile, who called a number of Arab and Jewish astronomers to his capital at Toledo for this express purpose. The Tables contained data for planetary positions and eclipse predictions; they appeared in 1521, and were used throughout Europe for the following three hundred years

URANUS

PROCYON B

The size of Procyon B compared with Uranus.

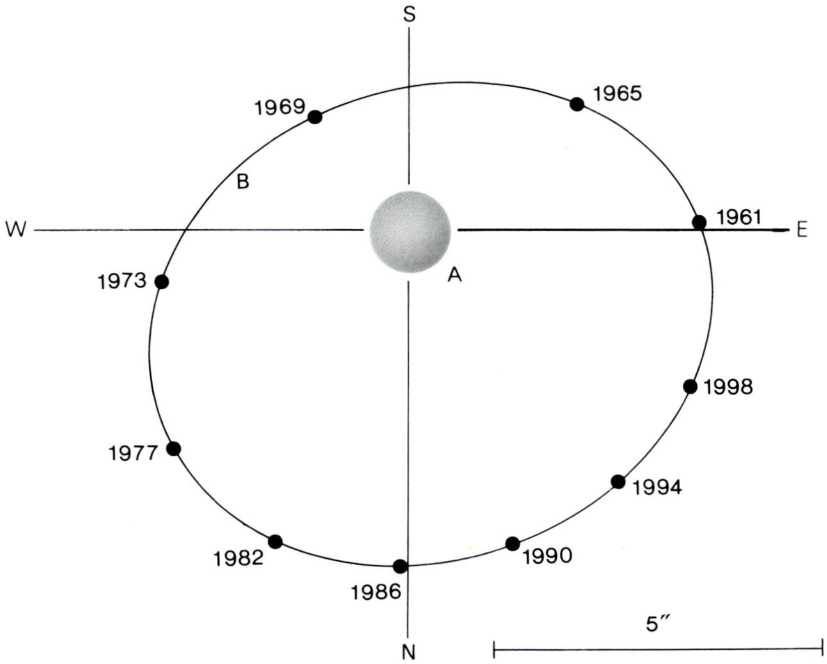

The apparent orbit of Procyon B.

8. PROCYON

Diagram of Canis Minor (Little Dog), showing Procyon.

Alpha Canis Minoris: 10 Canis Minoris

Right ascension:	07h 39m 18s.1
Declination:	+05° 13' 30"
Apparent magnitude:	0.38
Absolute magnitude:	+2.6
Spectral type:	F5
Luminosity, Sun=1:	7
Distance, light-years:	11.4
Parallax:	0".285
Radial velocity:	-3 km/s
Proper motion (per year):	RA 0".047, dec. -1".03

It is customary for a hunter to be attended by dogs. Orion has two. One is, of course, Canis Major, headed by the glittering Sirius. The other is Canis Minor, one of the smaller groups in the sky; it covers only 183 square degrees, and would be entirely unremarkable but for the presence of one brilliant star, Procyon. There is in fact only one other star above the fourth magnitude, Beta Canis Minoris or Gomeisa (2.9).

There are no special legends attached to Canis Minor. Procyon (formerly often spelled Procion or Prochion) was once referred to as Antecanus, Before the Dog, because it rises shortly before the Dog-star Sirius; to the Chinese it was Non Ho, the Southern River; in Babylonia it was Kakkab Paldaea, the Star of the Crossing of the Water-dog. Ulugh Beigh, last of the great astronomers of the Arab School, called it Al-Shamiyyah, and in the seventeenth century Riccioli knew it as Siair Siami; both these names may be interpreted as "the Little Sirius".

Astrologically it was associated with wealth and fame, and there was an interesting note by Leonard Digges in his book *Prognostica Everlasting*, published in 1553: "Who learned in matters astronomical, noteth not the great effects of the rising of the starre called the Litel Dogge?" Whether Digges believed in astrology is open to question; certainly it seems that he built the first telescope, more than half a century before its official invention in Holland.

Procyon is one of our nearest stellar neighbours; among first-magnitude stars, only Alpha Centauri and Sirius are closer. In itself we must admit that it is unremarkable, and were it not in our own part of the Galaxy it would be inconspicuous. True, it is more luminous than our Sun, and hotter; its diameter is probably around 3,500,000 miles (5,700,000 km), and it is a typical F-type Main Sequence star. Probably its main interest lies in its companion, which was discovered in a somewhat curious way.

Because of its closeness, Procyon has a relatively large proper motion, and, like Sirius, it seems to "weave" its way along against the background of more remote stars. As we have seen, Friedrich Bessel studied the motion of Sirius, and maintained that it was being pulled on by a much dimmer companion; after Bessel's death, the Companion, the Pup, turned up almost exactly where he had expected. It seemed logical to assume that Procyon had a companion of the same type, and in 1861 the German astronomer Georg Arthur Auwers made calculations indicating that the unknown companion moved round Procyon in a period of 40 years. Searches for it were made by several observers, notably S.W. Burnham in America and Otto Struve, who had succeeded his father Friedrich as Director of the Pulkova Observatory, in what is now Estonia and was then part of Russia. They had no success; the Companion remained elusive.

Next in the hunt was Simon Newcomb, of the observatory at Washington in the United States. He was waiting for the completion of a large new telescope — a 26-inch refractor made by Alvan Clark, whose reputation as an optical worker was second to none. Newcomb intended to use it to search for the companion to Procyon,

The Nearest Naked-Eye Stars

There are very few naked-eye stars within close range. Those which lie at less than 13 light-years are as follows:

Star	Apparent magnitude	Absolute magnitude	Spectrum	Luminosity Sun=1	Distance lt-yrs
Alpha Centauri	-0.27	4.4+5.7	G+K	1.6+0.45	4.4
Sirius	-1.46	1.4	A	26	8.7
Epsilon Eridani	3.73	6.1	K	0.30	10.8
61 Cygni	5.22	7.5+8.4	K+K	0.08+0.04	11.1
Epsilon Indi	4.68	7.0	K	0.14	11.2
Procyon	0.38	2.6	F5	7	11.4
Tau Ceti	3.50	5.7	G8	0.45	11.8

All our other close neighbours are very feeble dwarfs.

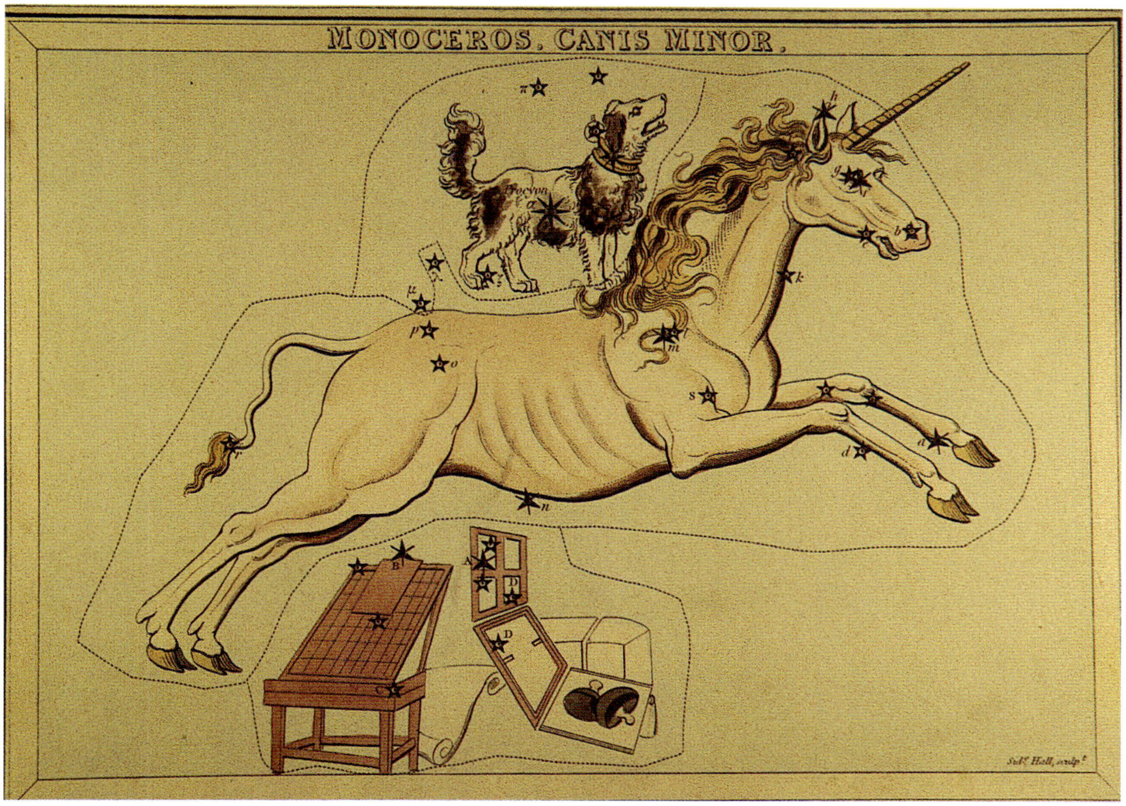

Old figures of Canis Minor and Monoceros, the Unicorn. Monoceros looks impressive here, but has no bright star!

and he wrote to Struve, at Pulkova, suggesting that a preliminary search might be worth while in spite of the earlier failures. Using the Pulkova 15-inch refractor in 1873, Struve announced that he had found the awaited star; the magnitude was given as 13, which seemed reasonable enough. It was also observed by Struve's assistants, E.E. Lindemann and A. Wagner, and accordingly Struve wrote back to Newcomb, asking him to provide confirmation.

By this time the Washington telescope was ready, and Newcomb lost no time in turning it toward Procyon. He fully expected to see the Companion, but there was absolutely no sign of it. He made many attempts, all of which were fruitless, and the whole situation was frankly puzzling, because Newcomb's telescope was much larger than Struve's and of equally good optical quality. In 1874 Struve visited London to address the Royal Astronomical Society, and reported on his detection of the Companion to Procyon. Still it remained invisible from Washington, and the mystery deepened.

At last Struve found the answer. He turned his telescope toward Regulus, Capella and Arcturus in turn, and, to his chagrin, all three showed faint companions similar to Procyon's. He realized that he was dealing with a telescopic ghost , a tiny imperfection in the optical system of his telescope, brought out by the light of any brilliant star.

The Christian Sky

In a star atlas published in 1627, Canis Minor is no longer a dog; it becomes the Paschal Lamb. The author of the atlas, a devout German Catholic lawyer named Schiller who lived in Augsburg, undertook a complete revision of the constellations, and attempted to Christianize them.

His scheme, given in his *Coelum Stellarum Christianum* (the Christian Celestial Heavens) was nothing if not revolutionary. For example Cassiopeia was transformed into Mary Magdalen; Hercules became the Three Magi; Centaurus changed into Abraham and Isaac, and Pegasus was no longer a flying horse, but the Archangel Gabriel. The twelve signs of the Zodiac became the twelve Apostles, while Draco, the Dragon, was changed into the Massacre of the Innocents. Schiller had connections with Johann Bayer, also an Augsburg lawyer and the man who first gave the stars their Greek letters, and the two planned a full revision of Bayer's *Uranometria*, with improved star positions. However, Schiller died in 1627 — the same year in which his atlas appeared — and, mercifully, his scheme was quietly forgotten.

However, the irregularities in the motion of Procyon were real enough, and it was hard to believe that no companion existed. It was finally found in 1896 by the American astronomer John Martin Schaeberle, using the 36-inch refractor at the Lick Observatory — apart from the Yerkes 40-inch, the largest refractor in the world (it still is). It proved to be very dim indeed, at about the 13th magnitude, and since it is never separated from its primary by much more than 5 seconds of arc it is an excessively difficult object. The apparent separation was greatest in 1990, and is now decreasing again toward a minimum of no more than 2 seconds of arc. The real separation is of the order of about 1400 million miles (2250 million km), which is rather less than the distance between our Sun and the planet Uranus.

Not unexpectedly, the companion proved to be a white dwarf; it is perhaps 17,000 miles (27,300 km) in diameter, slightly smaller than Uranus or Neptune, but with a mass about 65 per cent of that of the Sun, though it has only a tiny fraction of the Sun's luminosity. Its presence seems to make the existence of planets in the Procyon system very unlikely, though of course we cannot be sure.

New measures of the masses of the two components, based on their orbital motions, were made in 1994 by Canadian astronomers at the Dominion Observatory in British Columbia. They found that Procyon itself has 1.75 times the mass of the Sun; the value for the Companion was 0.6 that of the Sun. The surprise here is that if these figures are correct, Procyon ought to be more powerful than it actually is; it is underluminous by a factor of 2. Whether there is an error in the calculated mass, or whether there is some other explanation, remains to be seen.

Several faint stars lie nearby. Procyon C, at a distance of about 120 seconds of arc, is not a difficult object, but it lies well in the background, and has no connection with the Procyon pair. In 1833 Admiral W.H. Smyth, author of the classic *Cycle of Celestial Objects*, reported a 9th-magnitude star in the same telescopic field as Procyon; it was seen again in 1850, but not since, so that either Smyth was mistaken or else he just happened to catch a variable star or a nova when near its maximum.

Certainly Procyon lacks the splendour of Sirius or the majesty of Canopus or Rigel; but when looking at the Orion region, do not forget to take at least a passing look at the Little Dog.

Leonard Digges and his Telescope

Leonard Digges, who refers to Procyon in his book *Prognostication Everlasting*, was born in 1520, and had a distinctly chequered career. As a young man he narrowly escaped execution for his part in a rebellion against the reigning Queen of England, Mary Tudor. However, he was a good mathematician; he published mariner's almanacs, and *Prognostication* was crammed with scientific data.

He died in 1559, and left one son, Thomas, who re-issued *Prognostication* in 1576, adding an appendix of his own. In it there are clear indications that Leonard built a telescope at some time during the 1550s; the British historian Colin Ronan has collected very convincing evidence, though we have no idea what the telescope looked like or whether it was ever turned skyward. It seems to have used both a lens and a curved mirror, so that it was a curious device which must have been very difficult to use. We have no absolute proof that the telescope existed; but if it did, Digges anticipated the Dutch opticians by well over fifty years. The earliest telescopes which still exist were built around the year 1608.

The size of Achernar compared with the Sun.

Eridanus (The River).

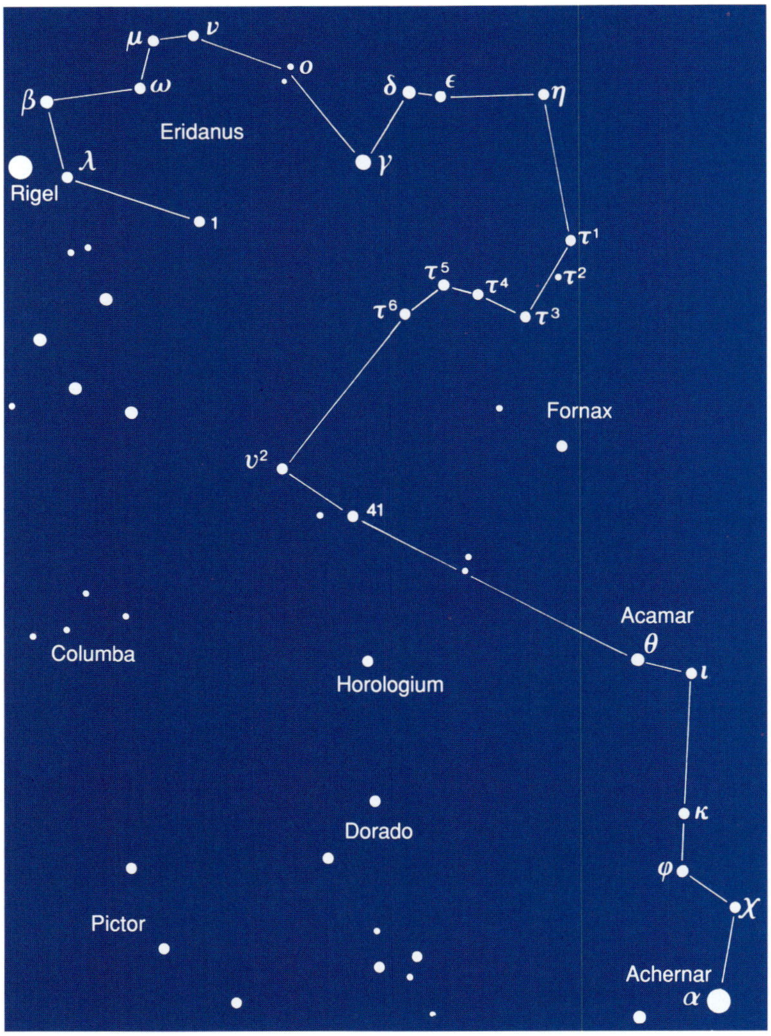

9. ACHERNAR

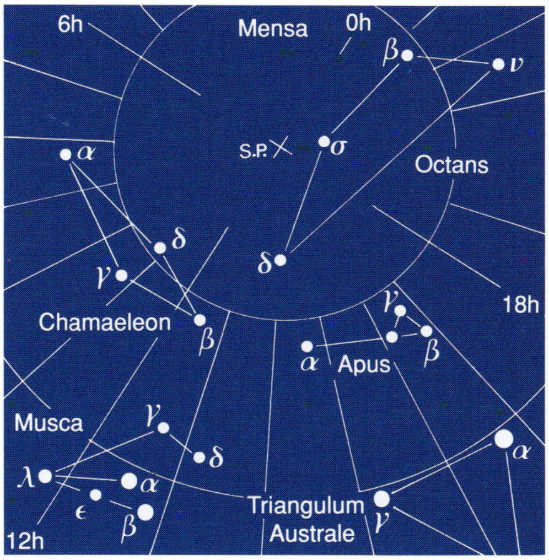

The position of the south celestial pole. The south pole star is Sigma (σ) Octantis.

Alpha Eridani

Right ascension:	01h 37m 42s.9
Declination:	-57° 14' 12"
Apparent magnitude:	0.46
Absolute magnitude:	-1.6
Spectral type:	B5
Luminosity, Sun+1:	400
Distance, light-years:	85
Parallax:	0".042
Radial velocity:	+19 km/s
Proper motion (per year):	RA +02".013, dec. -0".03

Achernar is the leader of the long, winding constellation of Eridanus, the River, which begins close to Rigel in Orion and ends in the far south. Though it is one of the largest constellations in the sky, covering well over 1000 square degrees, it contains remarkably few bright stars. Apart from Achernar, there are only three above the third magnitude; Theta (Acamar), Gamma (Zaurak) and Delta (Rana).

There are no really hard and fast legends associated with Eridanus, possibly because much of it, including Achernar, is invisible from any part of Europe. However, there are suggestions that it may be linked with the tale of Phaethon and the sun-chariot.

In this tale, Phaethon was the son of the sun-god, but his mother was mortal. When he reached manhood, Phaethon went to seek out his father, and was warmly greeted. Every day the sun-god Helios took the gleaming chariot across the sky, giving the world its light and warmth; the chariot was drawn by horses, and, rashly, Phaethon persuaded Helios to let him drive the chariot for one day. The result was disaster. The horses bolted; the chariot swooped down, and set the world on fire. Eventually the ruler of Olympus was compelled to strike Phaethon down with a thunderbolt, and he fell to his death

***Opposite** Star trails: the south celestial pole can be located, but there is no bright solar star. This photograph has a 10-minute time exposure.*
Photograph: Patrick Moore.

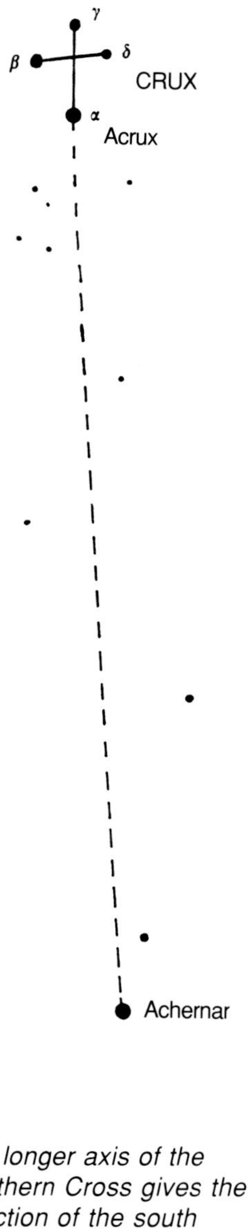

The longer axis of the Southern Cross gives the direction of the south celestial pole, beyond which is the brilliant Achernar. The pole is roughly midway between Achernar and the Cross.

in the river below — possibly Eridanus, though others have claimed that Eridanus is simply the Italian river Po.

Eridanus does contain some interesting objects, notably the star Epsilon, which is one of our nearest stellar neighbours and is regarded as a particularly promising candidate for a planetary centre. Here too is a notable dwarf pair, made up of Omicron^1 and Omicron^2 Eridani (Beid and Keid); Keid consists of a red dwarf accompanied by a white dwarf, which is an unusual combination. The red dwarf is of very low mass, and for a while was actually the least massive star known, though the record has now been broken by an even dimmer star, MH18, discovered by Michael Hawkins at Edinburgh from photographic plates taken in Australia. MH18 has only 1/20,000 the luminosity of the Sun.

Achernar, with its declination of 57 degrees south, is invisible from the whole of Europe and most of the United States, apart from Hawaii; it just skirts the horizon from Cairo, though it grazes the horizon even at its highest. It is circumpolar from Sydney and Cape Town, and of course from the whole of New Zealand.

There is a curious puzzle connected with it. The name Achernar may be translated from the Arabic, Al Anir Al Nahy, as "the End of the River", but originally was this Achernar at all? Ptolemy does not mention it, for obvious reasons, and the name seems then to have applied to a different star, Theta Eridani, which could be seen from Alexandria, where Ptolemy lived; its declination is 40 degrees south. Its name was first used in the famous Alphonsine Tables (though Ulug Beigh called it Al Thalim, the Ostrich). Nobody is quite sure which was the original Achernar, and to make matters even more confusing there have been suggestions that Theta has declined in brightness since ancient times. Ptolemy and Al-Sûfi both ranked it as of the first magnitude, but it is now only just above the third.

However, Acamar does not seem to be the kind of star which would undergo a permanent change in a few centuries. It is a fine double, with components of magnitudes 3½ and 4½ separated by over 8 seconds of arc, so that even a small telescope will show both components; the senior member is 50 times as luminous as the Sun, while the secondary could match 17 Suns. The distance from us is 55 light-years, and both stars are of type A, so that they would seem to be perfectly stable.

Come now to Achernar itself, which is the ninth brightest star in the sky. It is of spectral type B5, 85 light-years away and 400 times as luminous as the Sun. It is bluish-white, and of this there can be no doubt whatsoever; there are a few medieval reports describing it as red, but these are certainly due to the fact the observers looked at it when it was so low over the horizon that its light was strongly reddened. The surface temperature is about 14,000°C, and the diam-

A Planetary Centre?

Epsilon Eridani is far from prominent; its magnitude is only 3.7, but at least it is less than ten degrees south of the celestial equator, so that it can be seen from almost all inhabited countries. It lies not far from Rigel in Orion, and makes a pair with its neighbour Delta Eridani or Rana, which is slightly brighter (magnitude 3.5) and is a normal K-type star, 29 light-years away and three times as luminous as the Sun.

Epsilon is interesting because it is the nearest star which is reasonably like the Sun (it is slightly closer than Tau Ceti). Therefore it might well be expected to have a planetary system, and there are indications that it is being perturbed by a companion which is either a very massive planet or else a very low-mass star.

If we are to establish the existence of other intelligent beings, the only way which seems possible at the present moment is to pick up radio signals which are rhythmical enough to be interpreted as artificial. When "listening out" was first attempted in mid 1960, by astronomers using the radio telescope at Green Bank, West Virginia, Epsilon Eridani and Tau Ceti were the prime targets. Not surprisingly, the results were negative, but efforts continue. The chances of success may be small, but they are not nil. (*En passant*, the 600-ton Green Bank radio telescope suddenly collapsed on 15 November 1988, and was abruptly reduced to a heap of scrap metal. Mercifully, nobody was hurt.)

Of course, Epsilon Eridani is not a twin of our Sun. It is less luminous, with only 0.3 of the Sun's power, and has a K-type spectrum, so that it is cooler. Yet it is similar enough to be promising, and we cannot be sure that at this moment some radio astronomer in the Epsilon Eridani system is not doing his (or her, or its) best to pick up signals from the region of our Sun!

eter is probably about seven times that of the Sun — 6,000,000 miles, not far short of 10,000,000 km.

Achernar has the distinction of being the nearest really bright star to the south celestial pole, and it is in fact a help in locating the position of the polar point, which lies in a very barren area midway between Achernar and the Southern Cross (so that when Achernar is high up, the Cross is low down, and vice versa). There are few even reasonably bright stars closer to the pole; the nearest is Beta Hydri, in the Little Snake, which is of magnitude 2.8. However, even Beta Hydri is almost 13 degrees from the pole.

There is nothing really special about Achernar, but its rather isolated position makes it easy to locate, and it certainly does mark the End of the River, whether or not it was the star so named in ancient times.

Secular Variables

Acamar (Theta Eridani) is not the only star suspected to have undergone a permanent change in brightness in historic times; there are other alleged "secular variables", and in Eridanus there is also the case of Zeta Eridani, or Zibal. Both Ptolemy and Al-Sûfi ranked it of the third magnitude, but in more modern times Argelander classed it as 4, and it is now only 4.8. It is, in fact, rather strange that a star as obscure as Zeta Eridani now is should be given a proper name.

Another case is that of Denebola or Beta Leonis, in the Lion, which Ptolemy stated clearly was of the first magnitude, and equal to Regulus, the leader of the constellation. It is now just below magnitude 2. And in Ursa Major, the Great Bear, one of the seven "Plough" stars — Megrez, or Delta Ursae Majoris — is a magnitude fainter than any of the others, though some old reports made it equal to the rest.

However, it would be most unwise to place too much reliance on ancient records; it is only too easy to make a mistake in either interpretation or observation, and it is significant that none of the "classic" cases (Acamar, Denebola, Zibal or Megrez) seem to be stars which would be expected to change so markedly. We must certainly agree that at best, the claims are "non proven".

The South Celestial Pole

Southern navigators never cease to regret that there is no useful south polar star, comparable with Polaris in the northern sky. With the slightest mist or light pollution, the entire area will appear blank.

The polar constellation is Octans, the Octant; it was of course invisible to Ptolemy, and was added to the sky by the Abbé Lacaille in his maps drawn in 1752. The brightest star in Octans is Nu, of magnitude 3.8; the Bayer sequence of Greek letters has become wildly out of order here, and the magnitude of Alpha Octantis is only 5.2.

The nearest naked-eye star to the pole is Sigma Octantis, of magnitude 5.5. It is about 7 times as luminous as the Sun, with an A-type spectrum, so that in itself it is entirely unremarkable; moreover it is well over a degree from the actual pole, and precession is shifting it even further away. Its navigational use is obviously very limited. However, hope lies ahead. In the far future, when the northern hemisphere has Vega as its pole star, the south pole will have moved to a point not too far from the magnificent Canopus.

10. BETELGEUX

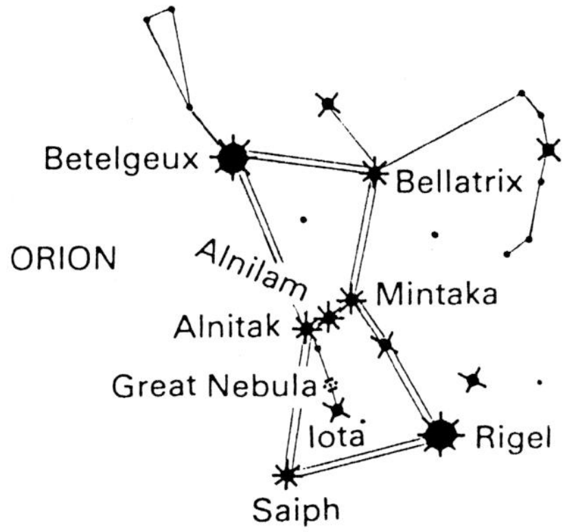

Diagram of Orion.

Alpha Orionis: 58 Orionis

Right ascension:	05h 55m 10s.2
Declination:	+07° 24' 26"
Apparent magnitude:	variable, 0.0–0.9
Absolute magnitude:	-5.6, variable.
Spectral type:	M2
Luminosity, Sun=1:	15,000
Distance, light-years:	310
Parallax:	0".007
Radial velocity:	+21 km/s
Proper motion (per year):	RA +0".002, dec.+0".01

Betelgeux is one of the most famous and most interesting stars in the sky. Though lettered Alpha Orionis, it is (usually) slightly below Rigel in brightness, but it is markedly variable, and at maximum has been known to reach zero magnitude, though generally it remains between 0.3 and 0.8.

The name is Arabic, and comes from Ibt-al-Jauza, the "Shoulder of the Central One", but it has been modified over the ages; in the Alphonsine Tables it is given as Beldengenze, and in the mid-17th century Riccioli called it Bedalgeuze. Today it can be spelled in various ways; Betelgeuse and Betelgeuze are variants. There are also different ways of pronouncing it, and some people make a habit of referring to it as "Beetlejuice"! It is spectacular by any standards, and it is hard to improve upon the description of it given a hundred years ago by the eminent English amateur astronomer William Lassell: "A most beautiful and brilliant gem! Singularly beautiful in colour, a rich topaz; in hue and brilliancy different from any other star I have seen."

There is absolutely no doubt about its changes in light; they were first noted by Sir John Herschel as long ago as 1836. The total magnitude ranges is less certain, but at best it is comparable with Rigel and Capella, while at minimum it is little, if at all, brighter than

The Garnet Star

Mu Cephei, in the far north of the sky, was nicknamed the Garnet Star by Sir William Herschel because of its fiery hue. Seen through binoculars, it has been compared with a glowing coal, but although it is redder than Betelgeux the colour is not evident with the naked eye because the star is so much fainter. It is variable, with a range of from magnitude 3.6 to 5.1; it has been classified as semi-regular, with a period of 430 days, but this period is very rough, and there are spells of irregularity.

Mu Cephei is about 1560 light-years away, and it has 53,000 times the luminosity of the Sun, so that it is very much more powerful than Betelgeux. Appearances can be highly deceptive!

Aldebaran in Taurus. Unfortunately, it is an awkward star to estimate by conventional naked-eye methods, because it has to be compared with another star which is not too unequal to it, and the only useful comparisons available are Capella (0.08), Rigel (0.12), Procyon (0.38) and Aldebaran (0.85). One has always to allow for the effects of extinction; that is to say, the dimming of a star's light when it is low down, and shining through a thicker layer of the Earth's atmosphere. It is therefore necessary to select a comparison star which is at about the same altitude, and this is none too easy. John Herschel recorded that he had occasionally seen it brighter than Rigel, and "actually the largest star in the northern hemisphere of the sky"; I have been following it for more than fifty years now, and at its brightest (as in November 1995) I have ranked it equal to Capella. On the other hand, I have never seen it fainter than Aldebaran, so that my personal opinion is that the range is between 0.0 and 0.9.

There is also the question of a possible period. It has been classed as irregular, but in the official catalogues it is given a rough period of 2110 days (that is to say, 5¾ years), and it is classified as a semi-regular variable. However, this period is very rough indeed, and is subject to marked fluctuations. The variations are intrinsic; Betelgeux swells and shrinks, changing its total output as it does so.

The surface temperature is much lower than that of the Sun, and is of the order of 3400°C; as with all M-type stars, its spectrum shows strong molecular bands, which are not present in stars with higher surface temperatures simply because the molecules would be broken up into their constituent atoms. According to the Cambridge catalogue, its distance is 310 light-years. Other catalogues increase this to 520 light-years, but in any case Betelgeux is definitely the nearest of the really vast red supergiants; this is why it is so conspicuous, though some other red naked-eye stars are much more powerful. Mu Cephei, nicknamed the Garnet Star, is a case in point.

The diameter of Betelgeux is thought to be about 250,000,000 miles (400,000,000 km), so that the huge globe could easily contain the orbit of the Earth round the Sun. However, it is much less dense than the Sun in its outer layers, because it is at a much later stage in its evolution. As its mass is between 15 and 20 times that of the Sun, it has run through its life-cycle much more quickly; it has used up its main store of hydrogen "fuel", and is drawing on its reserves, so that the core temperature must have rocketed to a staggering 100,000,000 degrees at least. Its outer layers are ten thousand times more rarefied than the Earth's air at sea-level, and the time will come when these outer layers will be puffed off altogether. Eventually, there is every chance that Betelgeux will explode as a supernova. If this happened, it would be an awe-inspiring spectacle in our skies, though the crisis may well be deferred for a million years or more.

The size of Betelgeux compared with the Sun.

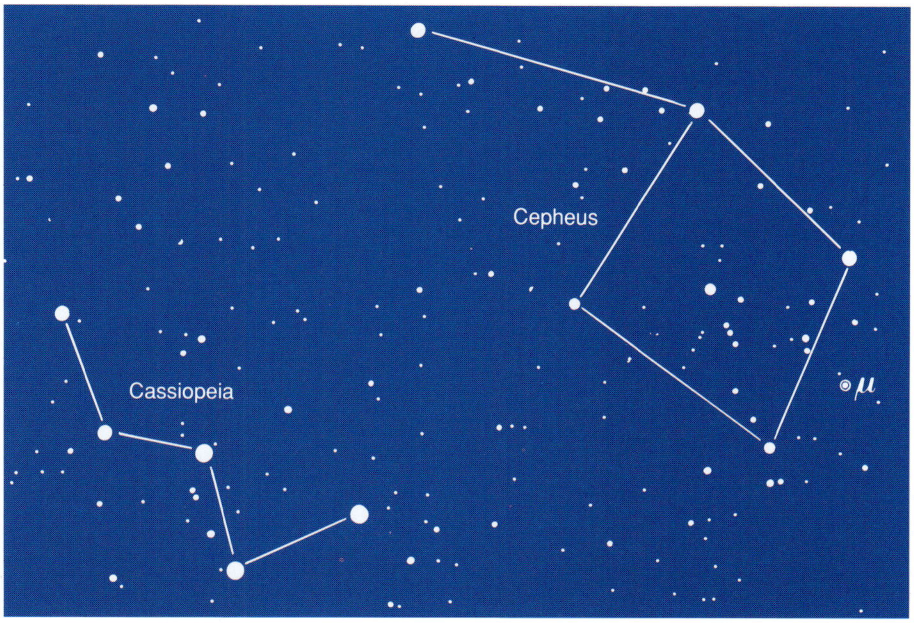

Cassiopeia and Cepheus showing the position of Mu Cephei.

Bright Orange or Red Stars

Stars of spectral type M are relatively cool, and are orange or orange-red. The following is a list of M-type stars above the fourth magnitude; of course, all are giants or supergiants.

Star		Name	Magnitude	Luminosity Sun=1	Distance, lt-yrs
β	Andromedae	Mirach	2.1	115	88
τ	Aquarii		4.0	110	440
λ	Aquarii		3.7	130	230
σ	Canis Majoris		3.5	15,000	7500
μ	Cephei		3.6(max)	52,500	1560
υ	Ceti		4.0	130	260
α	Ceti	Menkar	2.5	130	130
ο	Ceti	Mira	1.7(max)	var	95
γ	Crucis		1.6	160	88
λ	Draconis		3.8	110	210
τ^4	Eridani		3.7	130	235
η	Geminorum	Propus	3.1(max)	130	186
μ	Geminorum	Tejat	2.9	130	230
β	Gruis	Al Dhanab	2.1	170	750
α	Herculis	Rasalgethi	3.0(max)	700	220
γ	Hydri		3.2	120	160
σ	Librae	Zubenalgubi	3.3	132	166
α	Lyncis		3.1	120	166
δ	Ophiuchi	Yed Prior	2.7	130	140
α	Orionis	Betelgeux	0.0(max)	15,000	310
β	Pegasi	Scheat	2.3(max)	300	180
ρ	Persei	Gorgonea Terti	3.0(max)	130	196
L^2	Puppis		2.6(max)	1500	75
δ	Sagittae		3.8	700	550
η	Sagittarii		3.1	750	420
α	Scorpii	Antares	1.0	7500	330
μ	Ursae Majoris	Tania Australis	3.1	120	156
ν	Virginis		4.0	130	160
δ	Virginis	Minelauva	3.4	130	147

Ordinary telescopes show stars as mere points of light, and this means that stellar diameters are very difficult to measure by direct methods. Since Betelgeux is the nearest of the supergiants, it might be expected to have the largest apparent diameter, and in 1920 an attempt to measure it was made by A.A. Michelson and F.G. Pease, using the 100-inch Hooker reflector at Mount Wilson — then much the most powerful telescope in the world. They developed what is termed a beam interferometer, which produces interference fringes. It consists of two movable mirrors at the opposite ends of a steel girder which is placed across the open end of the telescope; the separation between the mirrors can be varied, in this case between 2 and 5.5 metres. Both mirrors receive the light from the star, and send it down to the main mirror at the bottom of the tube. The light from

Extinction

The closer a star is to the horizon, the more of its light will be lost. The following table gives the amount of dimming for various altitudes. Above an altitude of 45 degrees, extinction can be neglected for most practical purposes:

Altitude, degrees	Dimming, in magnitudes
1	3
2	2.5
4	2
10	1
13	0.8
15	0.7
17	0.6
21	0.4
26	0.3
32	0.2
43	0.1

There are additional problems in comparing a red star with a white one, and none of the useful comparisons of Betelgeux are orange or red apart from Aldebaran, which is of spectral type K.

the opposite sides of the tiny disk of the star arrives at slightly different angles, and the mirrors are moved until the two images are superimposed and the fringes disappear; the star's apparent diameter can then be worked out from the separation of the mirrors. In this way Michelson and Pease found that the apparent diameter of Betelgeux was 0.045 of a second of arc (the modern value is 0.047), making the real diameter about 30 times that of the Sun. They also measured other red stars; Antares gave an apparent diameter of

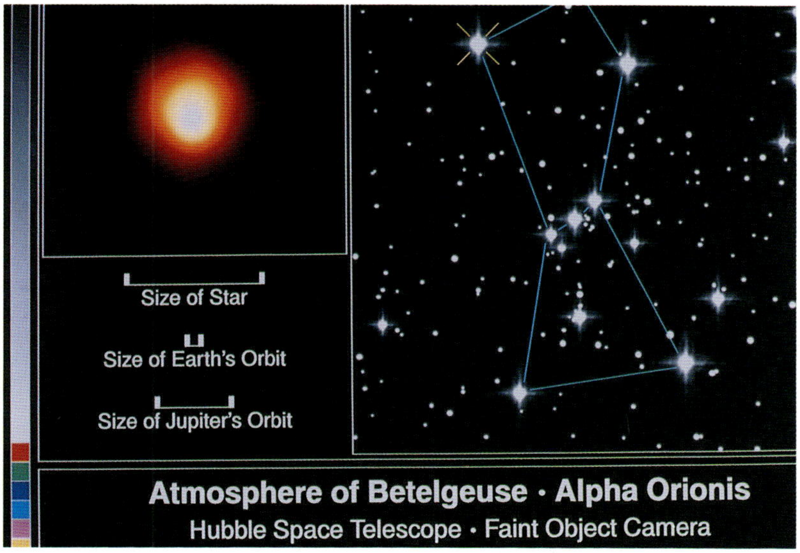

The atmosphere of Betelgeux — a picture from the Hubble Telescope.

The Rocks of the Boys, Los Muchachos. These rocks, atop the mountain, have given their name to the whole mountain itself. Los Muchachos is an extinct volcano — at least one hopes it is extinct! Certainly it is an excellent observing site and the summit is very often well above the top of the cloud layer.
Photograph: Patrick Moore

0.040, and Mira Ceti was slightly larger, at 0.056 of a second of arc. In fact Mira is not nearly so large as Betelgeux, but is less than 100 light-years away.

Recently a new system, developed by astronomers at the Sydney Observatory in Australia, is starting to yield much more precise values, and has confirmed that our estimate of Betelgeux is very near the truth. Of course, the diameter changes perceptibly as the star swells and contracts, and this has been checked spectroscopically.

Other new techniques, involving what is termed speckle interferometry, have enabled us to make a start in plotting surface features on Betelgeux. It seems that there is one "hot spot", a vast convective feature in the nature of a column of rising gas, which is responsible for about 10 per cent of the total light we receive. The main work here has been carried out by astronomers using the 165-inch William Herschel telescope on La Palma, in the Canary Isles.

Naturally, Betelgeux radiates over a wide range of wavelengths, and has been found to be a radio source; if our eyes were sensitive to infra-red, it would be one of the most brilliant objects in the entire sky. It is associated with a huge cloud of dust, and also a shell of expanding gas which may have been ejected around 5000 years ago and now extends out to a distance of over 9000 astronomical units from the star. Nearby there are also features termed molecular clouds, which are concentrations of dust and cold gas.

It is most unlikely that any planets orbit Betelgeux; certainly any close Earth-type worlds would have been destroyed when the star left the Main Sequence and moved over to the giant branch of the Hertzsprung-Russell diagram. If planets did exist, and were inhabited, astronomers there would see our Sun shining as a very dim star of the twelfth magnitude.

Sunset over the William Herschel Telescope dome at La Palma.
Photograph: Patrick Moore, 1994.

The William Herschel Telescope at La Palma.
Photograph: Patrick Moore, 1995

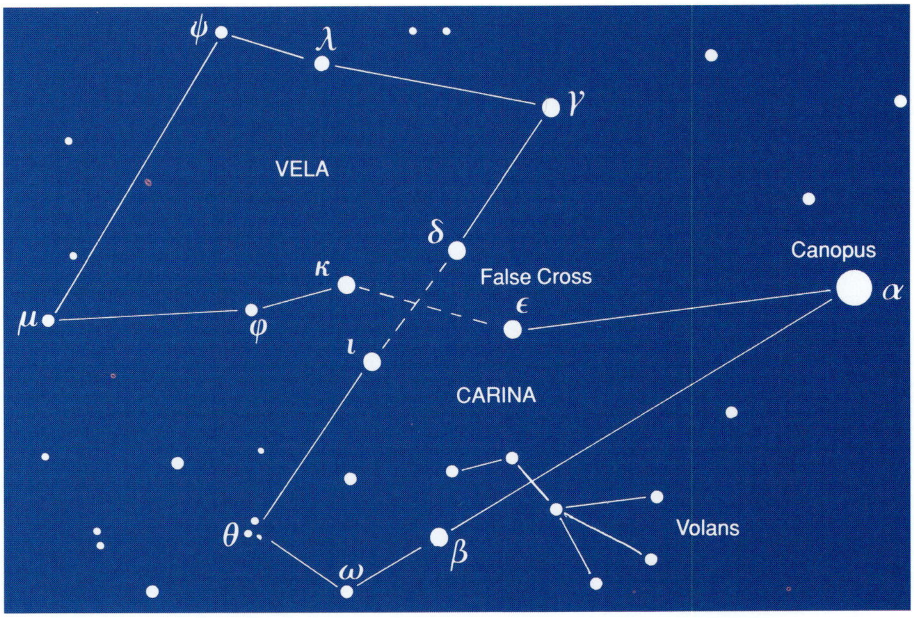

The False Cross.

Crux.
Photograph: John Dunlop, Auckland Observatory

11. AGENA

Beta Centauri

Right ascension:	14h 03m 49s.4
Declination:	-60° 22' 22"
Apparent magnitude:	0.61
Absolute magnitude:	-5.1
Spectral type:	B1
Luminosity, Sun=1:	10,000
Distance, light-years:	460
Parallax:	0".010
Radial velocity:	-11 km/s
Proper motion (per year):	RA -0".003, dec. -0".02

Beta Centauri, the second of the Pointers to the Southern Cross, has no real connection with its brighter neighbour. As we have seen, Alpha Centauri is the nearest brilliant star beyond the Sun, and is only slightly over 4 light-years away; Beta lies at 450 light-years, over a hundred times as remote. The Pointers make up a very

R Centauri

Close to Agena, more or less between it and Alpha Centauri, is a very red Mira-type variable, R Centauri. Its extreme range is from magnitude 5.1 to 13, so that when at its best it can be seen with the naked eye. It is of special note because the period seems to be shortening; it was 568 days early in the present century, but this has now decreased to 547 days. Moreover, it shows double maxima and minima. The minima alternate between magnitudes 9 and 11, with occasional falls to 13; the maxima are less unequal, but the slightly brighter maximum always follows the shallower minimum. We may here be watching a real change in the star's evolution. The distance is of the order of 1000 light-years, and the star is around 500 times as luminous as the Sun.

R Centauri is within binocular range for part of its period, and is always easy with a telescope. Its position is RA 14h 16m.6, declination -59º 44'; it is 1½º east-north-east of Agena. Its variability was discovered by B. Gould as long ago as 1871.

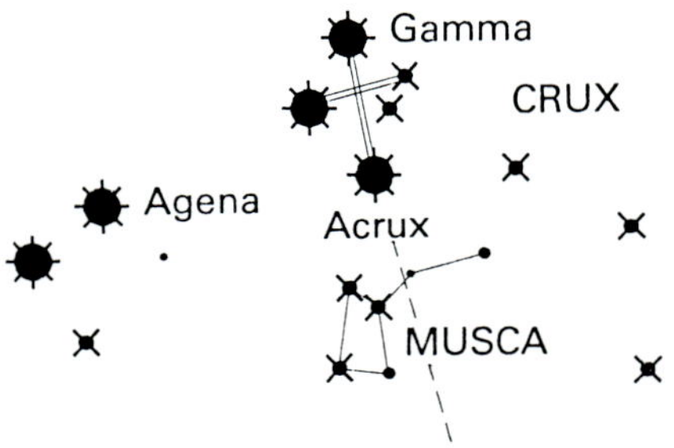

prominent pair; there are not many other pairs which can be regarded as comparable. Indeed, only Castor and Pollux come to mind, and the Twins are very much fainter than the Pointers.

The name Agena was first applied to Beta Centauri in the 17th century, but the origin of the name is rather obscure. The alternative is Hadar, often used by air and sea navigators; apparently the names Hadar and Wazn were formerly applied, somewhat indiscriminately, to both the Pointers. The Aborigines of Australia referred to them as the Two Brothers, while to the African Bushmen they were The Two Men who were Once Lions. The Chinese called Beta Mah Fuh, the Horse's Belly, since it lies in that part of the Centaur's anatomy.

With its declination of -30 degrees, Agena is not visible from anywhere in Europe; it skirts the horizon in the southern part of Florida, and is easy to see from Hawaii. From Australia and South Africa it

The Pointers: Alpha and Beta Centauri. The Pointers lie side by side, but they are quite unrelated.
Photograph: Patrick Moore

sets briefly, but from New Zealand it is always above the horizon.

There is nothing particularly notable about Agena itself, apart from its usefulness in showing the way to the Southern Cross. It is a typical bluish-white star of type B. It has a companion of magnitude 3.9, lying at a distance of 1.3 seconds of arc at a position angle of 251 degrees; this is not a difficult telescopic object, and would be very easy indeed were it not so overpowered by the primary. There is no doubt that the two are genuinely associated, but they must take many centuries to complete an orbit round their common centre of gravity, since they are at least 200 astronomical units apart. The companion is over 400 times as luminous as the Sun.

Of all the first-magnitude stars only Canopus, Rigel, Betelgeux and Deneb are more powerful than Agena. If it were as close to us as Alpha Centauri, it would cast strong shadows.

Star Visibility

It may be useful to give the visibilities of various stars from different latitudes. C indicates that the star is circumpolar; V that it can be seen at some time or other; a dash that it cannot be seen at all. Borderline cases are given in brackets.

Calculation is easy, as can be shown by the following example. The declination of Agena is, in round figures, -60°. 90-60 = 30. Therefore, Agena is circumpolar as seen from any latitude south of 30°S, and cannot rise from any latitude north of 30°N.

NORTH	α UMI 89 Polaris	α UMa 62 Dubhe	η UMa 49 Alkaid	α Aur 46 Capella	α Boö 19 Arcturus	α Ori 7 Betelgeux	β Ori 8 Rigel	α CMa 17 Sirius	α Sco 26 Antares	α PsA 30 Fomalhaut	λ Sco 37 Shaula	α Car 53 Canopus	α Eri 57 Achernar	α Cen 61 Alpha Centauri	α Cru 63 Acrux	σ Oct 89 Sigma Octantis
Hammerfest, 71	C	C	C	C	(C)	V	V	(V)	-	-	-	-	-	-	-	-
Reykjavik, 64	C	C	C	C	V	V	V	V	-	-	-	-	-	-	-	-
St Petersburg, 59	C	C	C	C	V	V	V	V	V	(V)	-	-	-	-	-	-
Aberdeen, 57	C	C	C	C	V	V	V	V	V	(V)	-	-	-	-	-	
London, 51	C	C	C	C	V	V	V	V	V	V	(V)	-	-	-	-	-
Montreal, 45	C	C	C	(C)	V	V	V	V	V	V	V	-	-	-	-	-
Rome, 42	C	C	(C)	V	V	V	V	V	V	V	V	-	-	-	-	-
Athens/San Francisco, 38	C	C	V	V	V	V	V	V	V	V	V	-	-	-	-	-
Jerusalem, 32	C	C	V	V	V	V	V	V	V	V	V	V	(V)	-	-	-
Cairo, 30	C	C	V	V	V	V	V	V	V	V	V	V	V	-	-	-
Delhi/Cape Canaveral, 29	C	(C)	V	V	V	V	V	V	V	V	V	V	V	(V)	-	-
La Palma, 28	C	V	V	V	V	V	V	V	V	V	V	V	V	(V)	-	-
Hilo, 20	C	V	V	V	V	V	V	V	V	V	V	V	V	V	V	-
Manila, 15	C	V	V	V	V	V	V	V	V	V	V	V	V	V	V	-
Mombasa, 4	C	V	V	V	V	V	V	V	V	V	V	V	V	V	V	-
Singapore, 1	(C)	V	V	V	V	V	V	V	V	V	V	V	V	V	V	-

SOUTH	α UMI 89 Polaris	α UMa 62 Dubhe	η UMa 49 Alkaid	α Aur 46 Capella	α Boö 19 Arcturus	α Ori 7 Betelgeux	β Ori 8 Rigel	α CMa 17 Sirius	α Sco 26 Antares	α PsA 30 Fomalhaut	λ Sco 37 Shaula	α Car 53 Canopus	α Eri 57 Achernar	α Cen 61 Alpha Centauri	α Cru 63 Acrux	σ Oct 89 Sigma Octantis
Quito, 0.2	-	V	V	V	V	V	V	V	V	V	V	V	V	V	V	V
Darwin, 12	-	V	V	V	V	V	V	V	V	V	V	V	V	V	V	C
Fiji, 19	-	V	V	V	V	V	V	V	V	V	V	V	V	V	V	C
Rio de Janeiro, 23	-	V	V	V	V	V	V	V	V	V	V	V	V	V	V	C
Johannesburg, 26	-	(V)	V	V	V	V	V	V	V	V	V	V	V	V	V	C
Sydney/Montevideo/ Cape Town, 34	-	-	V	V	V	V	V	V	V	V	V	V	(C)	C	C	C
Auckland, 37	-	-	V	V	V	V	V	V	V	V	V	V	C	C	C	C
Wellington, 41	-	-	-	V	V	V	V	V	V	V	V	C	C	C	C	C
Invercargill, 46	-	-	-	(V)	V	V	V	V	V	V	V	C	C	C	C	C
Falkland, 51	-	-	-	-	V	V	V	V	V	V	V	C	C	C	C	C

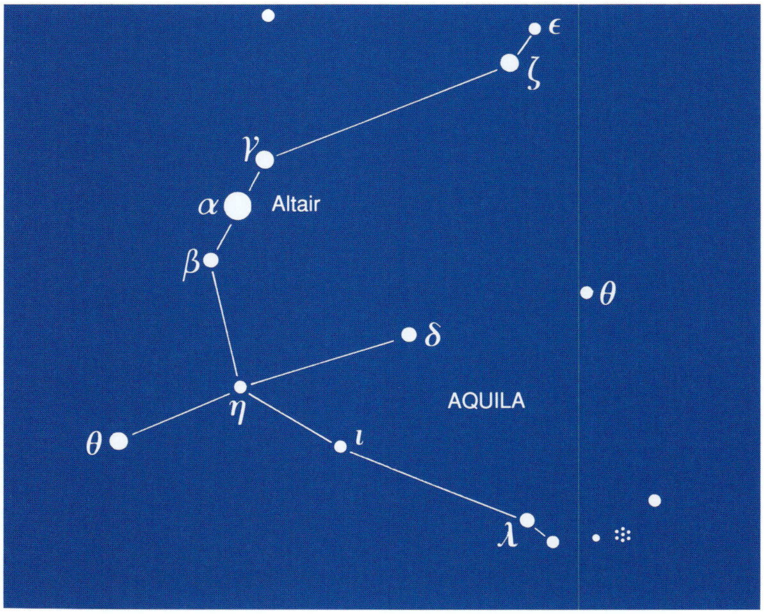

Aquila (The Eagle).

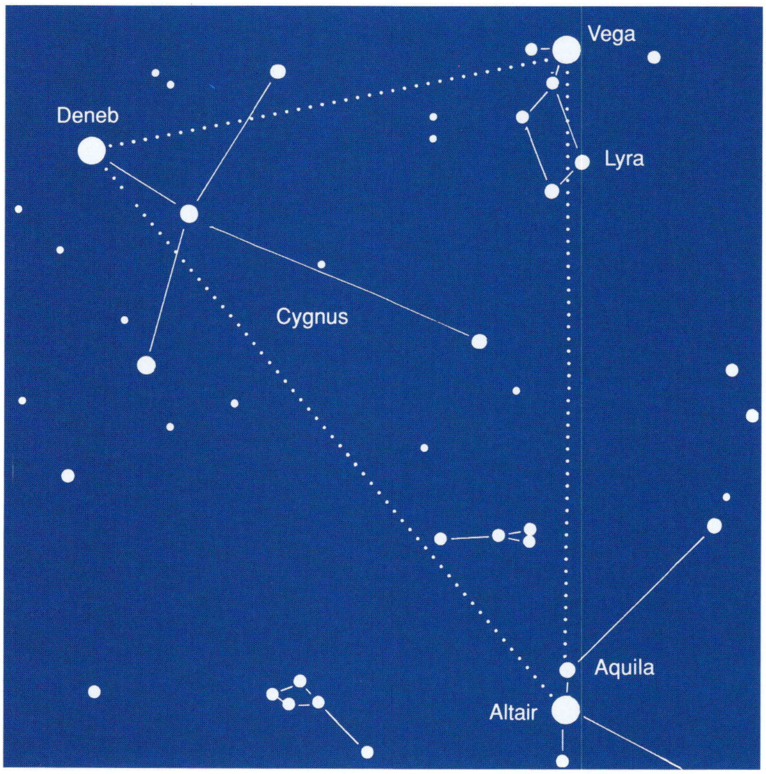

The Summer Triangle.

12. ALTAIR

Alpha Aquilae: 53 Aquilae

Right ascension:	19h 50m 46s.8
Declination:	+08° 52' 06"
Apparent magnitude:	0.77
Absolute magnitude:	2.2
Spectral type:	A7
Luminosity, Sun=1	10
Distance, light-years:	16.6
Parallax:	0".204
Radial velocity:	-26 km/s
Proper motion (per year):	RA +0".036, dec. +0".39

Altair is the leader of one of the most prominent of all constellations: Aquila, the Eagle. It is one of the few groups which may be said to give at least a vague impression of the object it is meant to represent; with sufficient imagination, one can conjure up the picture of an eagle in flight. Mythologically, it represents an eagle which was dispatched from Olympus to collect a shepherd-boy, Ganymede, who was destined to become the cup-bearer of the gods in succession to Hebe, who tripped and fell during a particularly solemn ceremony.

Most of Aquila lies in the northern hemisphere of the sky, though it does cross the equator. Altair has a declination of +16°, so that it can be seen from every inhabited country; only from a small part of Antarctica is it permanently out of view. Its name comes from part of the Arab name for the entire constellation, and has sometimes been written as "Althair" or "Atair". Astrologically it is regarded as a mischief-maker, and to give warning of danger from reptiles.

Altair makes up a large triangle with the first-magnitude stars Vega (in Lyra) and Deneb (in Cygnus). This has lead to a widespread nickname for which I admit I am responsible. In one of my *Sky at Night* television broadcasts, around 1958, I referred to these three stars as making up a "Summer Triangle", and this term has passed

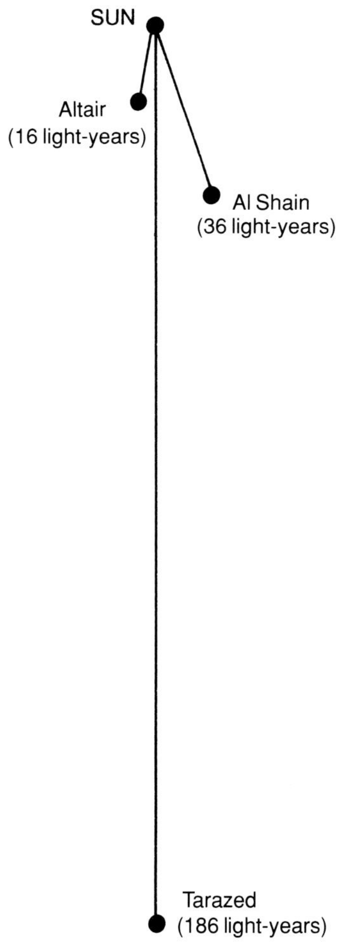

Distances from the Sun.

Old figure of Aquila, the Eagle. The Arrow (Sagitta) and the Dolphin (Delphinus) are also shown, with the rejected Antinoüs.

Antinoüs

Some old maps show a separate constellation, Antinoüs, which is made up of part of the modern Aquila, including the stars Eta, Theta, Delta and Lambda. It is said that the constellation was added to the sky on the orders of the Roman emperor Hadrian in honour of a boy who was very much a favourite of the Court. Whether this is true or not, Antinoüs never became generally accepted, though in his 1603 charts Bayer did show it and called it "Ganymedes". It has been dropped from all present-day star-maps.

into general use, though it is completely unofficial and the three members are in different constellations. Also, it does not apply to the southern hemisphere of the world, where the Triangle is at its best during winter; from the southernmost part of New Zealand Vega and Deneb barely rise at all.

There are two stars which flank Altair to either side; Beta or Alshain, and Gamma or Tarazed. Tarazed is a magnitude the brighter of the two, and is decidedly orange. The only other brilliant star to be the centre of a line of three is Antares in Scorpius, but there can be little danger of confusion, because Altair is pure white while Antares is fiery red.

Altair has a companion of magnitude 9.5, at a separation of 165 seconds of arc and a position angle of 301 degrees. However, there

The Royal Stars

The vernal or spring equinox is defined as the position in the sky where the Sun crosses the celestial equator, travelling from south to north; this happens each year in late March. The equinox now lies in Pisces, the Fishes, and is not marked by any bright star.

However, in former times the equinox lay in Aries (the Ram) and is still known as the First Point of Aries. It has been shifted because of the pull of the Sun and Moon on the Earth's equatorial bulge, producing what is termed precession. In the year 4000 BC the equinox lay in Taurus (the Bull), and it so happened that the equinoxes and the solstices were marked by four bright stars known as the Royal Stars: Aldebaran in Taurus (spring equinox), Regulus in Leo (summer solstice), Antares in Scorpius (autumnal equinox) and Altair (winter equinox). In fact Fomalhaut in Piscis Australis (the Southern Fish) would have been a more accurate choice, since the solstice actually lay in Aquarius, which is close to Fomalhaut; but in 4000 BC Fomalhaut was always very low in the sky as seen from latitude 40ºN, so that Altair was chosen instead.

is no real connection between the two; this is an optical double, not a binary system. The companion lies well in the background, while Altair itself is one of our closest stellar neighbours.

One interesting feature of Altair is that it is spinning very quickly on its axis. It completes one turn every 6½ hours, and this means that its globe is very distorted; the equator bulges out, so that Altair is elliptical in shape. The equatorial diameter is twice the polar diameter. Of course, something of the same effect is seen with all quick-spinning, non-rigid bodies; the flattening is very evident with

Alshain and Tarazed

Beta and Gamma Aquilae, flanking Altair, make an unequal pair. Though lettered Beta, Alshain is a magnitude the fainter of the two. It is of type G8, and slightly yellowish, though most observers will call it white; at any rate, it is very different from the orange Tarazed.

All the old observers, up to and including Johann Bayer in 1603, ranked Alshain as of the third magnitude, and equal to Tarazed. In 1890 the French astronomer Camille Flammarion went so far as to claim that "it is certain that it has diminished in brightness", but, as with all suspected secular variables, the evidence is very inconclusive at best.

Tarazed is much the more remote and more luminous of the two, and is indeed much more powerful than Altair. The details are as follows:

Tarazed:	mag. 2.72	distance 186 lt-yrs	type K3	luminosity 700 Sun
Alshain:	3.7	36	G8	4.5

If Tarazed were as close to us as Altair, it would shine in the sky as brilliantly as Venus.

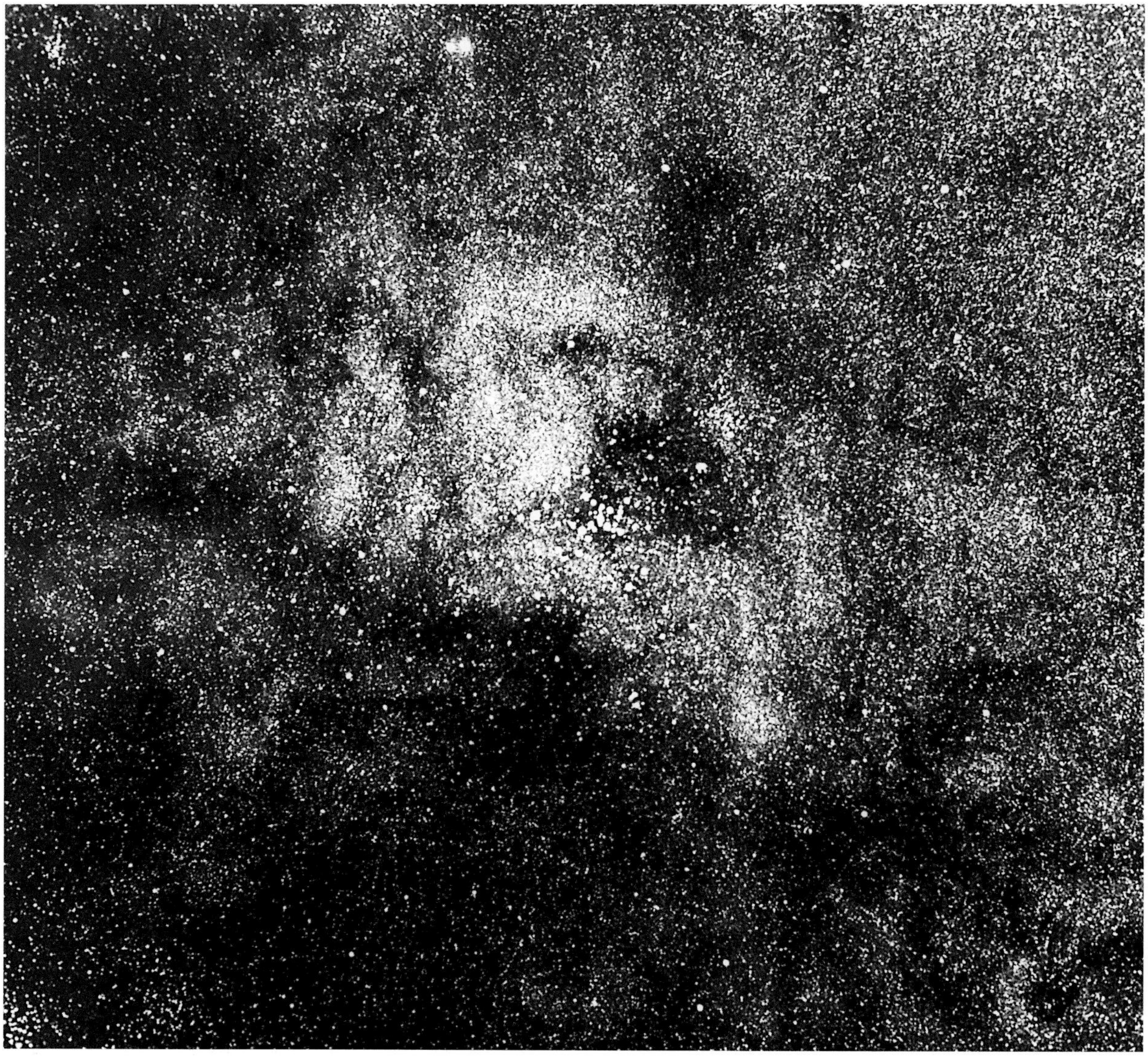

Rich star-field in the Milky Way.

the planets Jupiter and Saturn, for example — but Altair is an extreme case.

Since a star appears virtually as a point source of light, it may be asked how we can determine the rotation periods. As so often in astronomy, we rely upon the spectroscope. Altair has a spectrum of type A, with the usual absorption lines, but the lines are considerably broadened, because of the well-known Doppler effect. As a star spins, one of its limbs is approaching us while the other is receding (assuming that we have a broadside-on view, as with Altair; we must also take the overall radial velocity into account — at the moment Altair is approaching us at 16 miles (26 km) per second. Therefore, the approaching limb will show a blue shift, and with the receding limb the shift will be to the red. The effect is to broaden the lines, and the amount of the broadening gives a clue as to the rate of rotation.

It does not seem very likely that Altair is the centre of the planetary system, though we cannot definitely discount it. If an inhabited planet exists, astronomers there will have a curious view of their egg-shaped sun!

Eta Aquilae

South of Altair lies a line of three stars: Theta, Eta and Delta Aquilae. Theta and Delta are of magnitude 3.2 and 3.3 respectively; Eta is a Cepheid variable. Its magnitude range is from 3.5 to 4.4, so that it is always an easy naked-eye object, and, like all Cepheids, it is very regular, with a period of 7.2 days.

The first Cepheid to be detected has given its name to the entire class: Delta Cephei in the far north of the sky. The fluctuations were first noted by John Goodricke, in 1784. (Goodricke was a most unusual astronomer; he was deaf and dumb, and died at the age of 21, but there was certainly nothing wrong with his brain.) Other Cepheids were soon found, with periods ranging from a few days to a few weeks, and all were as regular as clockwork. Much later, it was found that the real luminosity of a Cepheid depends upon its period; the longer the period, the more powerful the star — and Cepheids are highly luminous; Eta Aquilae is the equal of 5000 Suns.

Once the period of a Cepheid has been measured, its luminosity can be found, and this gives us its distance, so that Cepheids act as invaluable "standard candles" in space. It was by discovering short-period variable stars in the Andromeda and Triangulum Spirals, in the 1920s, that Edwin Hubble was able to prove that the so-called "spiral nebulae" are independent galaxies rather than minor features of the Milky Way.

Only a few classical Cepheids exceed magnitude 4 at maximum:

Eta Aquilae:	range 3.5 to 4.4	period 7.2 days.
ZZ Carinae:	3.3 to 4.2	36.5
Delta Cephei:	3.5 to 4.4	5.4
Beta Doradûs	4.7 to 4.1	9.1
Zeta Geminorum	3.7 to 4.1	10.1

Of these, ZZ Carinae and Beta Doradûs lie in the far south of the sky, and never rise over Europe.

13. ACRUX

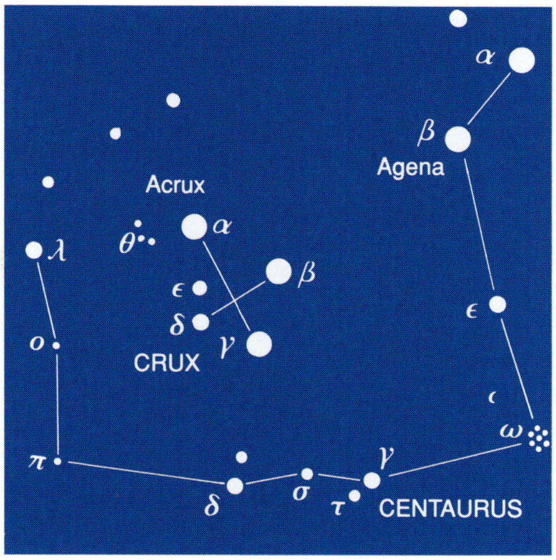

The Southern Cross and Pointers.

Alpha Crucis

Right ascension:	12h 26m 35s.9 (α^1)
Declination:	-63° 05' 56" (α^2)
Apparent magnitude:	0.83
Absolute magnitude:	-3.9 (α^1), -3.4 (α^2)
Spectral type:	B1 (α^1), B3 (α^2)
Luminosity, Sun=1:	3200 (α^1), 2000 (α^2)
Distance, light-years:	360
Parallax:	0".001
Radial velocity:	-11 km/s (α^1), -1 km/s (α^2)
Proper motion (per annum):	RA -0".004, dec. -0".02

Opposite *Crux and the Pointers.*
Photograph: John Dunlop, Auckland Observatory

Just as Ursa Major is the most famous of the northern constellations, so the Southern Cross, Crux Australis, is supreme in the south. Rather surprisingly, it is the smallest of all the accepted 88 constellations, with an area covering a mere 68 square degrees; it was not even listed separately until 1679, since before that date it was included in Centaurus, which surrounds it almost completely. Apparently it was first granted independence by an otherwise obscure astronomer named Augustin Royer, though it had been drawn clearly enough on earlier star-charts.

It is too far south to be seen from any part of Europe (about 3000 BC it did rise briefly from Alexandria, but the effects of precession mean that it no longer does so). However, it is wrong to assume, as many people do, that it is invisible from any northern country. The declination of Alpha Crucis, its brightest star, is -63 degrees. 90 minus 63 equals 27; therefore, in theory Alpha Crucis can rise above the horizon from anywhere south of latitude 27°N.

To be honest, Crux itself is nothing like a cross, and more nearly resembles a kite-pattern; there is no central star to make an X, as there is with Cygnus in the north. Moreover, the four main stars are unequal in brightness, and one of them, Delta Crucis, is more than a magnitude fainter than any of the others; there is also a fifth star,

Star Names

Generally speaking, proper names of stars are used for only the stars of the first magnitude and a few special cases, such as Mira (Omicron Ceti, the famous variable) and Mizar (Zeta Ursae Majoris, the naked-eye double in the Great Bear). Most of the accepted names are Arabic, though a few, such as Sirius, come from the Greek.

The name of Acrux was used in some older star-charts, but is neither Greek nor Arabic, and seems to have been "made up". Many people, even astronomers, prefer to refer to it simply as Alpha Crucis.

Epsilon Crucis, which spoils the symmetry. All the same, the constellation is so striking that it cannot be overlooked, particularly when taken together with the brilliant Pointers, Alpha and Beta Centauri.

Obviously there are no classical legends attached to Crux, but in some parts of Australia it was once said to represent a sting-ray being chased by a shark, while other Aborigine tribes said that it represented a white gum tree in which two yellow-crested cockatoos (the Pointers) were trying to roost! In parts of Black Africa it represented the Tree of Life, the holiest of all symbols, put into the sky to guide travellers who would otherwise lose their way.

The great explorer Alexander von Humboldt, writing in 1799, made some apt comments about Alpha and Gamma Crucis. He wrote:

"These two great stars, which mark the summit and foot of the Cross, having nearly the same right ascension, it follows that the constellation is almost perpendicular at the moment when it passes the meridian. This circumstance is known to the people of every nation situated beyond the Tropics or in the southern hemisphere. It has been observed at what hour of the night, in different seasons, the Cross is erect or inclined. It is a timepiece, which advances very regularly nearly 4 minutes a day, and no other group of stars affords to the naked eye an observation of time so easily made. How often have we heard our guides exclaim in the savannahs of Venezuela and in the desert, extending from Lima to Truxillo, 'Midnight is when Cross begins to bend'." He also commented that it was strange that Crux should not have been earlier recognized as a separate constellation, "especially since Kazwini and other Muhammadan astronomers took pains to discover crosses elsewhere in the sky".

Alpha Crucis is one of the most splendid binaries in the sky. Its double nature was detected in 1685 by a rather unexpected astrono-

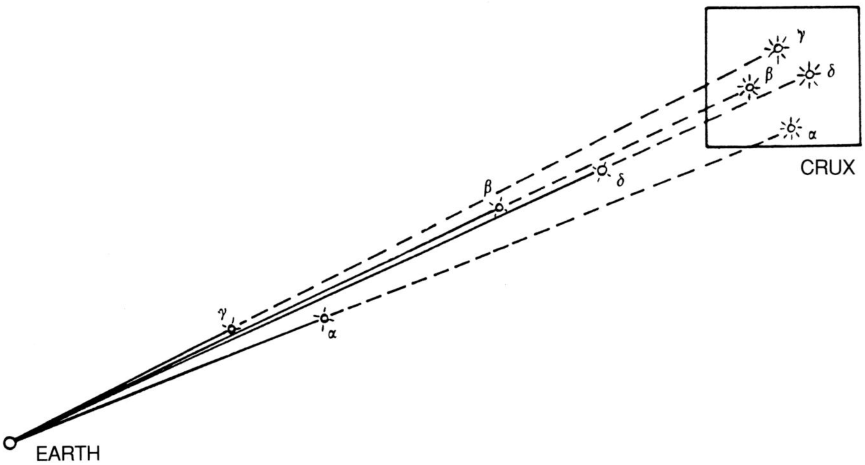

The stars in the Southern Cross are not genuinely associated. Gamma (γ) is the nearest; then comes Alpha (α), followed by Beta (β) and finally Delta (δ). They seem close together in the sky only because they happen to lie in much the same direction as seen from Earth. Delta is much further away from Gamma than we are.

Old star chart.

mer, the French Jesuit priest Father Guy Tachard. He and his party were on their way to Siam, and en route they called in at the Cape of Good Hope, which was then controlled by the Dutch. The Governor, Simon van der Stel, made his visitors very welcome, and Tachard proceeded to set up a temporary observatory with the aim of determining the longitude of the Cape, which was not then known with any accuracy at all. His method was to use his tiny telescope to time the eclipses and occultations of the four major satellites of Jupiter.

Humboldt and the Southern Cross

Alexander von Humboldt first saw the Cross in 1799, from a sailing boat in the southern Atlantic Ocean. In his diary he gives a graphic description of it:

"For several days the lower regions of the air had been filled with vapour, but on the night of the 4th to the 5th of July in 16 degrees latitude we saw the Southern Cross clearly for the first time. It was steeply inclined, and appeared from time to time between the clouds, the centre of which in the flickering sheet lightning shone with silvery radiance. If a traveller be permitted to mention his personal feelings, I may remark that on this night one of the dreams of my earliest youth were fulfilled."

He also made other observations, and in particular he looked at the Southern Cross, noting that "The foot of the Crozier marked in Bayer is a Double Star, that is to say, consisting of two bright Stars distant from one another about their own Diameter only, much like to the most Northern of the Twins; not to speak of a third much less, which also is to be seen, but further from these two."

The "foot of the Crozier" is, of course, Alpha Crucis; the northern Twin is Castor in Gemini, which was already known to be double. The "third star" is Alpha Crucis C, of magnitude 4.9, in the same telescopic field as the bright pair; its separation is about 90 seconds of arc. Almost certainly it is associated with the main pair, but it is a very long way from them.

Both components of the binary ($Alpha^1$ and $Alpha^2$) are of spectral type B, so that they are hot and bluish-white. Since the separation is over 4 seconds of arc, almost any telescope will split them.

The Two Crosses

It is rather a pity that the Southern Cross is asymmetrical, and that one of the four members of the pattern (Delta) is so much fainter than the other three. Epsilon Crucis, between Alpha and Delta, is a magnitude fainter again. It is interesting to look at their absolute magnitudes — that is to say, the apparent magnitudes which they would have if they could be seen from the standard distance of 10 parsecs (32.6 light-years).

Star	Apparent magnitude	Absolute magnitude	Spectrum	Distance, lt-yrs
Alpha	0.83	-3.9, -3.4	B1, B3	360
Beta	1.25	-5.0	B0	425
Gamma	1.63	-0.5	M3	88
Delta	2.80	-3.0	B2	260

Even a casual glance will show that Gamma Crucis differs from the others inasmuch as it is red. Some distance away is a similar pattern, the so-called False Cross, which lies partly in Carina (the Keel) and Vela (the Sails). Here the four members are:

Star	Apparent magnitude	Absolute magnitude	Spectrum	Distance, lt-yrs
Epsilon Carinae	1.86	-2.1	K0	200
Delta Velorum	1.96	+0.6	A0	68
Iota Carinae	2.25	-4.7	F0	250
Kappa Velorum	2.50	-3.0	B2	390

There is often confusion between the Crosses, particularly as in each case one member is obviously red (in the case of the False Cross, Epsilon Carinae). However, the False Cross is larger and less brilliant, and its stars are less unequal in magnitude. Carina and Vela are, of course, parts of the dismembered Argo Navis, the Ship Argo.

The Southern Cross taken from Bali, where the Cross is well seen. Note the differing colours of the stars. This was an unguided photograph with an exposure of between one and two minutes, to give slight star trails.
Photograph: Patrick Moore

They must take many centuries to complete one orbit round their common centre of gravity, and there is little obvious change in their position angle and separation even over many decades. Probably the real distance between them is at least 500 astronomical units, not far short of 50,000 million miles (80,000 million km).

Alpha^1 is obviously brighter than its companion. Both have been suspected of being spectroscopic binaries, and the spectral lines of Alpha^2 are broadened, so that if it is not a close binary it is presumably a quick spinner.

Crux is crossed by the Milky Way, so that the whole area is very rich. There can be few people of the southern hemisphere who can fail to recognize the Cross, particularly as it is circumpolar over much of Australia and South Africa and the whole of New Zealand.

Diagram of Taurus showing Aldebaran.

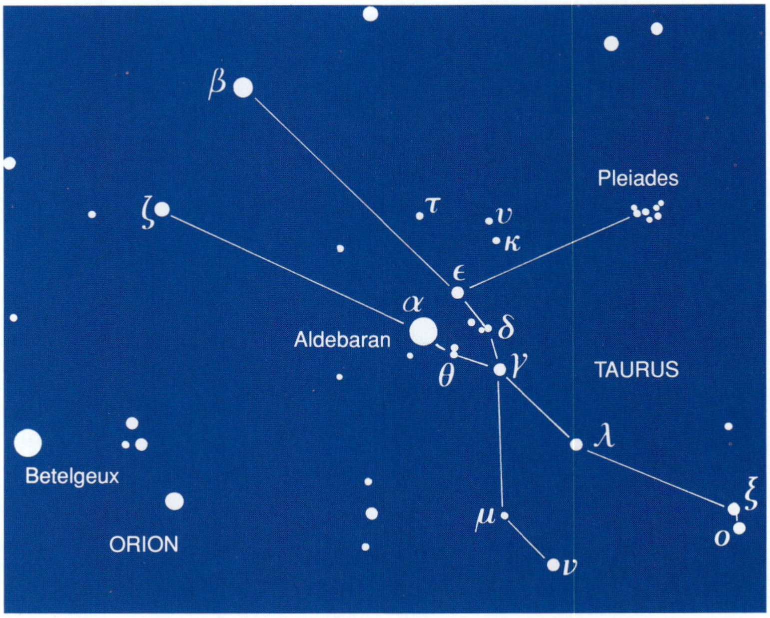

The Pleiades.

14. ALDEBARAN

Alpha Tauri: 87 Tauri

Right ascension:	04h 35m 55s.2
Declination:	+16° 30' 33"
Apparent magnitude:	0.85
Absolute magnitude:	-0.3
Spectral type:	K5
Luminosity, Sun=1:	120
Distance, light-years:	68
Parallax:	0".051
Radial velocity:	+54 km/sec
Proper motion (per year):	RA +0".005, dec. -0".19

Aldebaran is the leader of Taurus, the Bull, one of the most important of the Zodiacal constellations. Four thousand years ago it contained the vernal equinox, though since then the equinox has shifted right across Aries and into Pisces.

Taurus is a very old-established constellation; in mythology it is said to represent the bull into which Jupiter or Zeus, the king of the gods, changed himself in order to carry off Europa, daughter of the King of Crete. (The morals of the ancient Olympians were, to put it mildly, questionable!) The constellation cannot be said to have any particular shape, even though it has filched one of its main stars, Al Nath, from the adjacent constellation of Auriga, so that Gamma Aurigae has become Beta Tauri. However, Taurus contains the two most famous open star-clusters in the sky, the Pleiades and the Hyades, as well as the Crab Nebula, which is the remnant of the supernova of 1054, and which radiates over practically the whole range of the electromagnetic spectrum, so that it is of paramount importance to astrophysicists.

Aldebaran, the orange-red "Eye of the Bull", is easy to find. The name comes from Al Babaran, the "Follower" of the Pleiades; it is one of the Royal Stars, and in astrology was said to be fortunate, bringing wealth and riches. In the Alphonsine Tables it was Cor Tauri, the Bull's Heart; a Hindu name for it was Rohini, the Red Deer, doubtless because of its noticeable colour.

It lies in line with the three stars of Orion's Belt, and since it is less than 17°N of the celestial equator it can be seen from every inhabited country. It is one of the few first-magnitude stars sufficiently close to the ecliptic to be hidden or occulted by the Moon (the others are Spica, Regulus and Antares), and one of these occultations, recorded from Athens in March 509, led to an important discovery many years later.

Edmond Halley, best remembered for his association with the famous comet which bears his name, was England's second Astronomer Royal; he held office from 1721 till his death in 1742. He made some calculations, and found that Aldebaran could not possibly have

The Caldwell Catalogue

The French astronomer Charles Messier was interested mainly in comets, and discovered a number of them. During his searches, carried out with a small telescope, he was persistently misled by objects which looked cometary, but which turned out to be clusters or nebulae. Eventually, in 1781, he lost patience, and drew up a list of 109 nebulous objects to which he could refer without wasting time. Ironically, few people now remember his comets, but everyone uses his catalogue of clusters and nebulae.

He did not list the Hyades, because there was no danger of confusing them with a comet. However, he also omitted many other objects which are well within the range of modest equipment. In 1995 I decided to draw up a supplementary catalogue, also of 109 objects (all of which, naturally, are contained in the NGC or New General Catalogue, drawn up by the Danish astronomer J.L.E. Dreyer more than a century ago now). Since Messier and Moore both began with M, I could not use M numbers; but my surname is actually hyphenated — Caldwell-Moore — so I used the letter C. The Caldwell Catalogue was published in *Sky and Telescope* for December 1995, and has quickly become surprisingly popular!

Typical entries are:

C	NGC	Constellation	Type	Name
11	7635	Cassiopeia	Bright nebula	Bubble Nebula
14	869/884	Perseus	Double cluster	Sword-Handle
20	7000	Cygnus	Bright nebula	North America Nebula
31	IC 405	Auriga	Bright nebula	Flaming Star Nebula
41	-	Taurus	Open cluster	Hyades
49	2237-9	Monoceros	Bright nebula	Rosette Nebula
53	3115	Sextans	Galaxy	Spindle Galaxy
55	7009	Aquarius	Planetary nebula	Saturn Nebula
63	7293	Aquarius	Planetary nebula	Helix Nebula
77	5128	Centaurus	Radio galaxy	Centaurus A
80	5139	Centaurus	Globular cluster	Omega Centauri
92	3372	Carina	Bright nebula	Eta Carinae Nebula
93	6752	Pavo	Globular cluster	Pavo Globular
94	4755	Crux	Open cluster	Jewel Box (Kappa Crucis)
99	-	Crux	Dark nebula	Coal Sack
103	2070	Dorado	Bright nebula	Tarantula Nebula
106	104	Tucana	Globular cluster	47 Tucanae

Old figure of Taurus, the Bull.

been occulted in 509 if it had been in the position it occupies now. In fact, Aldebaran had shifted against the background of more distant stars, and has moved by about 7 minutes of arc, which is one-quarter of the apparent diameter of the Moon. Halley found that a few other bright stars, notably Sirius and Arcturus, had also moved in historic times, and this was the first discussion of the "proper" or individual movements of stars. Of course, proper motions are very slight, and if we could go back to the time of, say, Julius Caesar we would see the constellations in virtually their present form, but over a sufficiently long period the patterns will change, and if we could return in a few tens of thousands of years the sky-patterns would seem very unfamiliar.

Aldebaran is an orange-red giant. With the naked eye it looks very much like Betelgeux in Orion, but there are major differences between the two. Aldebaran is not nearly so large or so powerful as Betelgeux, and is "only" 120 times as luminous as the Sun; its diameter is of the order of 35,000,000 miles (56,000,000 km). Like so many orange and red stars it is slightly variable, but not to the same extent as Betelgeux; the range is only 2/10 of a magnitude (0.75 to 0.95), which is not enough to be noted with the naked eye, particularly as there are no really suitable comparison stars to hand. However, Aldebaran itself makes a good comparison star for Betelgeux. Gen-

Charles Messier, who drew up his famous Catalogue in 1781.

erally, Betelgeux is considerably the brighter of the two.

There are two faint companions. One is of the 13th magnitude, 31.4 seconds of arc from Aldebaran at a position angle of 112 degrees. It is not easy to see, because it is so overpowered by the brilliant orange glare of Aldebaran; it appears to be a red dwarf, with a luminosity less than 1/1000 of that of the Sun. It shares Aldebaran's motion through space, and is about the same distance from us, so that there appears to be a genuine association. The second companion is rather brighter, at magnitude 11; it is 121 seconds of arc from Aldebaran at a position angle of 034 degrees, and is a close binary made up of two red dwarfs. However, it is not connected with Aldebaran, and the apparent separation between the two is gradually increasing, so that here we have nothing more significant than an optical double.

Aldebaran seems to lie in the Hyades star-cluster, but appearances are misleading, because it is not a true cluster member. Its distance from us is 68 light-years, while the Hyades are of the order of 150 light-years away. Aldebaran lies in the foreground, about midway between the cluster and ourselves. In a way this is rather a pity, because the cluster is virtually drowned in the orange light, but the compact V-formation of stars is distinctive enough. Probably the best way to view them is through binoculars, since with a telescope the field of view will be too small to contain the whole cluster at once.

Altogether the Hyades cluster contains many dozens of stars, and there are at least 130 above the ninth magnitude. Excluding Aldebaran, the leaders are:

Star	Apparent magnitude	Absolute magnitude	Spectral type
Theta^2	3.42	0.5	A7
Epsilon (Ain)	3.54	0.2	K0
Gamma (Hyadum Primus)	3.63	0.2	K0
Delta^1	3.76	0.2	K0
Theta^1	3.85	0.2	K0
90	4.27	2.1	A5
Delta^3	4.30	1.4	A2
Upsilon	4.28	1.1	F0
Rho	4.65	2.6	F0
Sigma^2	4.68	2.1	A5
Delta^2	4.80	2.4	A7

Epsilon and Gamma are clearly orange. Delta^1 makes up a pair with the fainter Delta^2, while Theta is a naked-eye double, consisting of a white primary and an orange secondary; with binoculars, the colour difference between the two Thetas is striking. In fact they are

The Pleiades

The Pleiades, or Seven Sisters, make up the most celebrated of all open clusters, and legends about them come from almost every country. People with average eyesight can see at least seven stars without optical aid, and keen-eyed observers can exceed a dozen. The Pleiades are much younger than the Hyades, so that their leading stars are hot and bluish-white; they have not yet had time to evolve off the Main Sequence to the giant branch of the Hertzsprung-Russell Diagram. There is also considerable associated nebulosity, easy to photograph though rather elusive visually.

The brightest Pleiad is Alcyone, of magnitude 2.9. Next in order come Electra, Atlas, Merope, Maia, Taygete, Pleione, Celaeno and Asterope. The cluster is No. 45 in Messier's list.

The Pleiades. Note the nebulosity — easy to photograph but strangely difficult to see visually.
200-inch photograph, Palomar

not true close neighbours, since the white star is some 15 light-years nearer to us, and has been found to be a spectroscopic binary with a revolution period of 141 days.

Little or no interstellar material exists in the cluster, so that star formation has ceased there. Moreover, none of the leaders are hot bluish stars of type B. The Hyades cluster is relatively old, so that the main stars have evolved more than the principal stars of the younger Pleiades. Many low-luminosity stars have been identified

Legends of the Hyades

According to Greek mythology, the Hyades were the daughters of Atlas and Aethra. When their brother Hyas was killed by a wild boar, the sisters were so grief-stricken, and wept so copiously, that the gods took pity on them and transferred them to the sky, which is why the cluster is associated with stormy and wet weather. To the Romans they were often referred to as the Sucluae, or Little Pigs, though there seems no obvious reason for this, and it has alternatively been suggested that the name indicates Sucus, or moisture.

Certainly they have been recorded since very ancient times, and they are mentioned by both Homer and Hesoid.

— feeble red dwarfs, and ancient white dwarfs; as seen from the Hyades, our Sun would shine as a dim object not much above the eighth magnitude.

The Hyades make up what is termed a moving cluster. The whole group is drifting toward a convergent point in the sky not far from Betelgeux; perspective effects mean that over the ages they will seem to draw closer together. They were nearest to us about 800,000 years ago, and since the overall diameter of the cluster is at least 80 light-years, they would then have seemed much more scattered than they do now. Fifty million years hence, the Hyades will appear as a dim cluster only 20 minutes of arc across, and Aldebaran will long since have ceased to be a pseudo-member.

The Hyades.

15. ANTARES

Alpha Scorpii : 21 Scorpii

Right ascension:	16h 29m 24s.3
Declination:	-26° 25" 55"
Apparent magnitude:	0.96
Absolute magnitude:	-4.4
Spectral type:	M1
Luminosity, Sun+1:	7500
Distance, light-years:	330
Parallax:	0".009
Radial velocity:	-3 km/s
Proper motion (per year):	RA 0".000, dec. 0".02

Of all the constellations of the Zodiac, it is perhaps fair to claim that the Scorpion is the most magnificent. It is always rather low from Britain and most of mainland Europe, but when seen near the zenith, as it is from southern countries, it is truly spectacular. It is also one of the few constellations to conjure up at least a vague impression of the object after which it is named; the long curved line of stars, of which Antares is the most brilliant, can be said to resemble the shape of a scorpion. It is crossed by the Milky Way, and is one of the richest parts of the entire sky.

Astrologically it is of often referred to as "Scorpio", but Scorpius is the correct form. Antares is the 15th brightest star in the sky; also in the constellation are Lambda Scorpii or Shaula (24th) and Theta or Sargas (40th), which are not far below the first magnitude.

In early times Scorpius also included the stars of the constellation we now call Libra, the Balance or Scales, its neighbour in the Zodiac. Originally Libra was "Chelae Scorpii", the Scorpion's Claws, and indeed the star which used to be known as Gamma Scorpii has been transferred to Libra, as Sigma Librae (one of several similar cases; we have already noted the transfer of Al Nath from Auriga to Taurus, while Delta Pegasi, in the Square of Pegasus, has become Alpha Andromedae). There is, however, a minor mystery here. Early

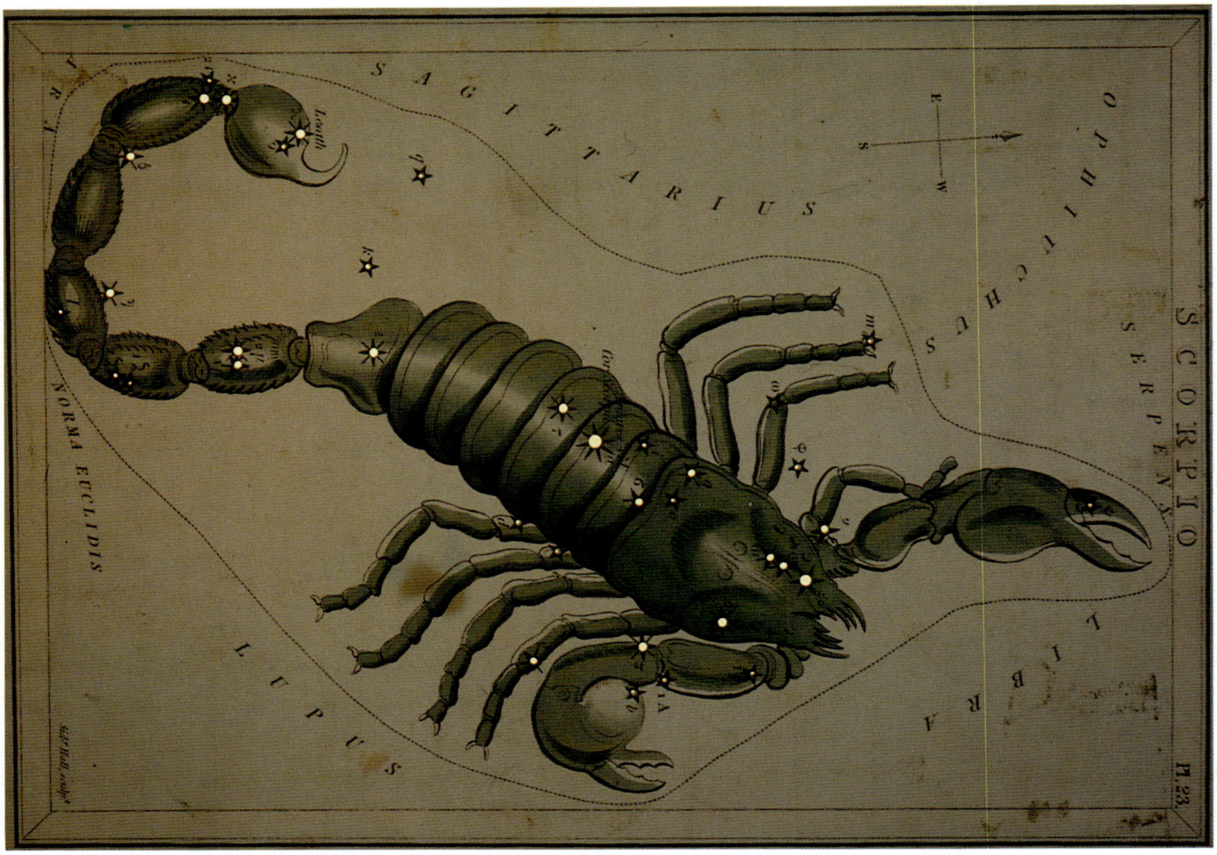

Old figure of Scorpius — placed in the sky as far away from Orion as possible!

observers ranked Antares as equal or inferior to Beta Librae or Zubenelchemale, the "Northern Claw", as was clearly stated around the year 350 BC by one of the great Greek scientists, Eratosthenes. He wrote:

"Due to its great size the Scorpion was divided into two signs: in the one, the claws remained, in the other, the body and the stinger; each claw has two stars, one brilliant and the other dark; three more brilliant stars are located up front, two at its belly, five at its tail, four at the stinger. These stars are headed by the most beautiful of them all, the very brilliant one of the Northern Claw."

This is definite enough, and around AD 150 Ptolemy, last of the great astronomers of Classical times, also made Beta Librae at least equal to Antares. Yet today, Beta is only a magnitude 2.6, which makes it a full 1½ magnitudes inferior to Antares. Certainly this is strange, because it does not seem likely that Beta Librae has decreased in brightness; it is a B8-type star, just over 100 times as luminous as the Sun, and would not be expected to show a permanent change over a few centuries. It is equally unlikely that Antares has increased, so that again we seem to be dealing with a mistake in observation or (more probably) in interpretation.

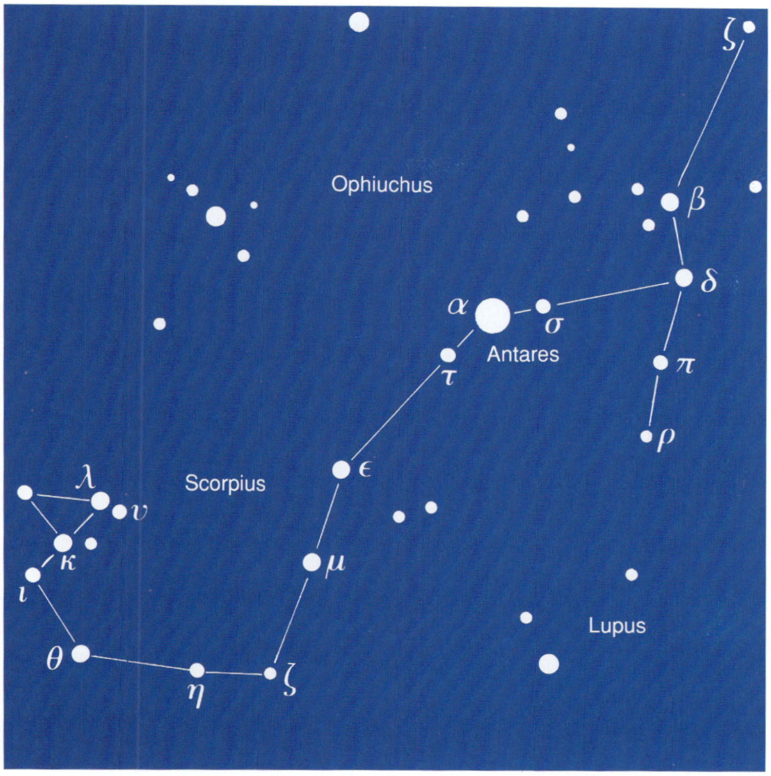

Left *Diagram of Scorpius.*
Below *Photograph of Scorpius.*
Patrick Moore

Green Stars

Orange and red stars are common; there are many yellow stars (including the Sun) and others are decidedly bluish, but single green stars are to all intents and purposes non-existent. In fact, the only single star commonly said to have a decidedly greenish hue is Beta Librae, and even here most observers will certainly call it white. The reported greenish tinge of a few other stars (Zeta Coronae Borealis, Delta Herculis and Delta Cygni, for example) is even more elusive.

There are, however, some double stars in which the fainter component is described as green. In each case, as with Antares, the primary is orange or red. Certainly the companion of the red supergiant Alpha Herculis (Rasalgethi) looks green by contrast. The same is true of the companion of Gamma Delphini, in the little constellation of the Dolphin; the primary is a yellowish G-type star (magnitude 4.5) while the companion is a magnitude fainter. The separation is almost 10 seconds of arc, so that the pair is much easier to split than Antares.

In Classical mythology, Scorpius is linked with the story of Orion; it alone of all earthly creatures could kill the great hunter — which it did, by stinging Orion in the heel. When both were placed in the sky, they were prudently set as far apart as possible. To the Chinese, Scorpius was part of the mighty Azure Dragon, residence of the Blue Emperor, while Antares was the Fire Star; in the Dendereh Zodiac of Ancient Egypt, we see the Scorpion much as it is portrayed today in astrological almanacs. Another Egyptian legend tells how Horus, son of the deities Osiris and Isis, was killed by a scorpion, but was brought back to life by his mother — and certainly there was always a link between the constellation and Isis.

Antares itself lies well south of the celestial equator, but it can be seen from most inhabited countries, though it is lost from North Norway and just touches the horizon from Reykjavík in Iceland. It is at its best from May to July, but from countries such as Australia

Attendants of Antares

Like Altair in Aquila, Antares is flanked to either side by a fainter star — in this case Sigma Scorpii or Alniyat (magnitude 2.9) and Tau Scorpii (2.8), which has never been given an individual name. Both are bluish-white stars of type B, and both are almost as luminous as Antares; they are of course further away. If they could be seen from our standard distance of 32.6 light-years, all three would shine as brightly as Venus does to us, and would make a magnificent trio. Even the companion of Antares would be of magnitude 0.5.

South Polar star trails.

and New Zealand it is well on view for a substantial part of each year.

Though Antares is by no means the equal of Betelgeux, it is a huge red supergiant. To the Romans it was Cor Scorpii (the Heart of the Scorpion), but the name we use comes from "Ant-Ares", the Rival of Mars, Ares being the Greek equivalent of Mars; the name is logical enough, since Antares is one of the reddest of all the bright stars, and its colour is striking even at first glance. When Mars passes close to it in the sky, it is interesting to compare the two. When Mars is of about the first magnitude it looks very like Antares, and it is strange to remember that Mars is a planet smaller than the Earth, shining only by reflected light, while Antares is well over 7000 times as powerful as the Sun.

Antares is one of the few stars whose diameter can be measured directly. The apparent diameter is only slightly less than that of

Johann Bürg

The discoverer of the companion to Antares was primarily an observer of the Moon, and it was because he was studying the lunar motions that he paid close attention to occultations. He was born in 1766, and became Professor of Astronomy and Mathematics at the University of Vienna; his lunar tables, published in 1806, were more accurate than any previously produced. He retired in 1819, and died in 1834. A crater on the Moon has been named in his honour.

A southern hemisphere view of Scorpius and Sagittarius setting in the west, with Jupiter below Antares.
Photograph: John Dunlop, Auckland Observatory

Betelgeux, though of course Antares is considerably the closer of the two; the real diameter must be of the order of 200,000,000 miles (320,000,000 km), as great as that of the Earth's orbit. However, the density is low at least in the outer layers, and the mass is probably no more than ten times that of the Sun. Like most red supergiants it is slightly variable, but the magnitude range (0.86 to 1.06) is too slight to be noticed with the naked eye.

There is a binary companion of magnitude 5.4, which would be an easy object were it not so drowned by the glare of the primary; even so, it is not hard to see with a modest telescope. It was discovered by the Austrian astronomer Johann Bürg, from Vienna, on 13

April 1819. Antares had been occulted by the Moon (as we have noted, it is one of the few first-magnitude stars sufficiently close to the ecliptic to be occulted), and when it emerged from behind the Moon it was the companion which appeared first. The revolution period is 878 years, and at present the apparent separation is decreasing; it was 3.1 seconds of arc in 1940, but is now only 2.6 seconds, and in the early 22nd century it will be hard to see the two stars separately. The real distance between the two is of the order of 500 astronomical units.

The Companion is of spectral type B4, but it is only 50 times as powerful as the Sun, so that it is decidedly underluminous for its class. It is usually described as green in colour. In part this may be due to contrast with the red primary, but it has been noted that the green hue is evident even when the star is emerging from occultation and for a brief period can be seen shining on its own. Whether or not this is accurate is somewhat dubious, but during occultations the effect is worth checking.

The outer atmospheres of stars show continuous expansion, and this radial outflow is termed "stellar wind". One obvious result is a loss of mass — inappreciable for mild stars such as the Sun, but very marked for supergiants such as Antares. The Companion is orbiting in a region strongly affected by the stellar wind from the supergiant, and this is one reason why the system emits radio waves. The radio emissions are of two types: one coming from the weakly ionized inner portions of the wind, and the other from localized regions where the wind has been "lit up" by the Companion. Other supergiants with binary attendants show the same effect, but it is particularly marked with Antares.

The whole area is enveloped in bright and dark nebulosity, and is well worth sweeping with binoculars or a low-power, wide-field telescope. Certainly the celestial Scorpion has much to offer, and this would still be the case even if it were not graced by the presence of Antares.

Messier 4

In the same binocular field as Antares lies a prominent globular cluster, No. 4 in Messier's catalogue. It is on the fringe of naked-eye visibility, and has been known for a long time; it was discovered by the French astronomer de Chéseaux in 1746. It is a fairly "open" globular, and the outer parts can be resolved with a modest telescope. It contains many thousands of stars; the true diameter is almost 100 light-years, and the distance from us has been given as 7500 light-years, rather close at hand by globular cluster standards. It is only one and a half degrees west of Antares.

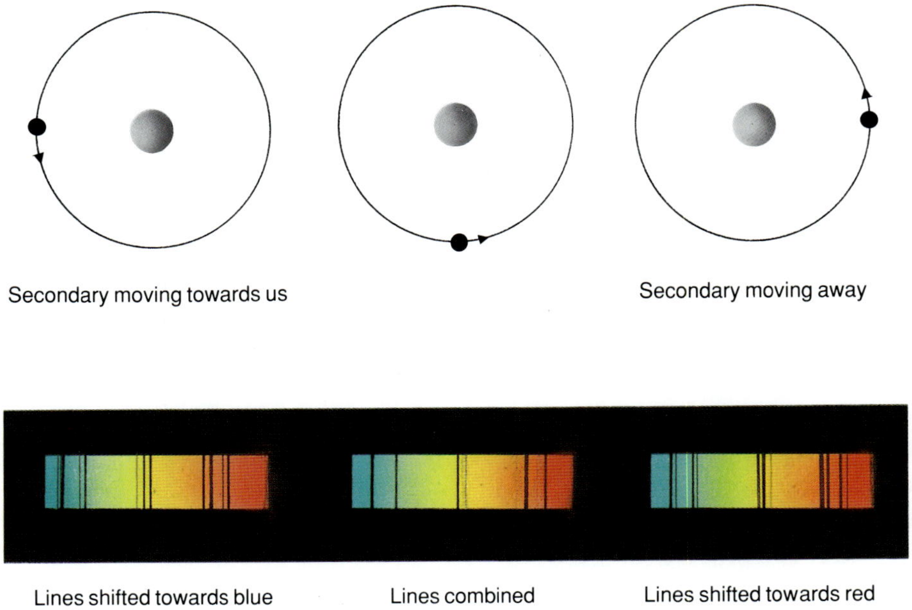

The Spectroscopic Binary Principle.

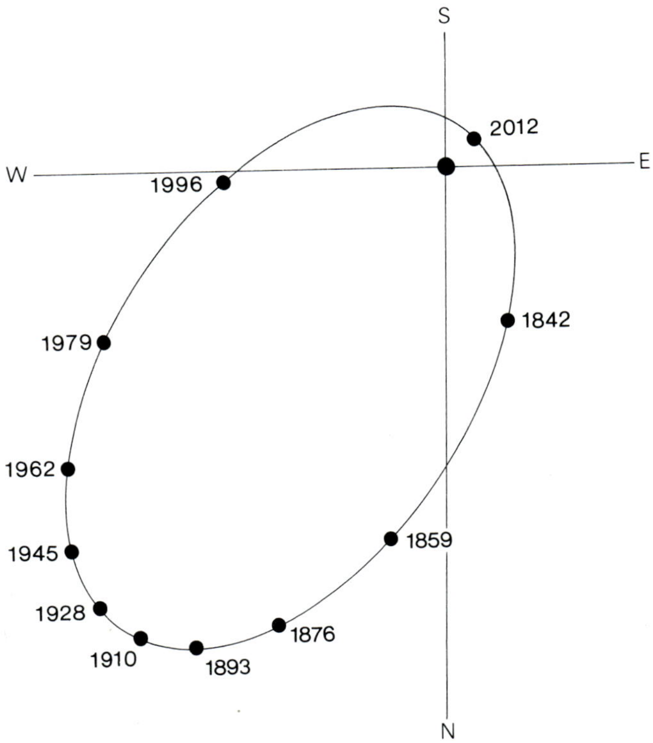

The changing separation of Gamma Virginis (based on one component remaining fixed).

16. SPICA

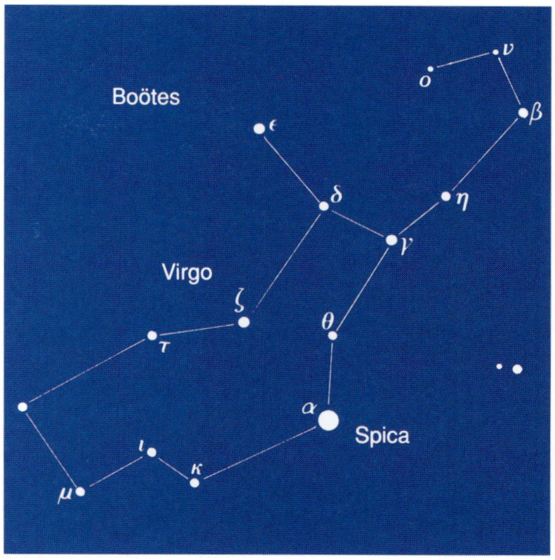

Virgo.

Alpha Virginis: 67 Virginis

Right ascension	13h 25m 11s.5
Declination:	-18° 18' 21"
Apparent magnitude:	0.98
Absolute magnitude:	-3.5
Spectral type:	B1
Luminosity, Sun=1:	2100
Distance, light-years	260
Parallax:	0".015
Radial velocity:	+1 km/s
Proper motion (per year):	RA -0".003, dec. -0".03

Virgo is the largest of the Zodiacal constellations, and now that Argo has been so ruthlessly chopped up it is actually the largest constellation in the sky apart from Hydra; it covers 1294 square degrees. Though it contains only one first-magnitude star, Spica, it includes a wealth of important telescopic objects.

It is very easy to locate when Ursa Major is on view. Simply follow round the three stars of the Bear's "tail" (Alioth, Megrez and Alkaid); this line, somewhat curved, leads first to Arcturus in Boötes and then to Spica. If Ursa Major is not available, it may help to remember that Spica forms a large equilateral triangle with Antares and the Southern Cross. During evenings it is at its best from May to July.

Most legends associate Virgo with a woman carrying a sheaf of wheat, but there are diverse opinions about just who she represents. She is generally take to be Astraea, the goddess of wisdom, daughter of Jupiter (Zeus) and Thetis, but alternatively she may be Proserpina, daughter of the earth-goddess Ceres, who was waylaid by Pluto and now spends six months of every year in the Underworld. In Egypt, Virgo was associated with the goddess Isis, while in the Chinese zodiac she was the Woman in the Ship.

Spica's name indicates "the Ear of Wheat", and there are suggestions that various Greek and Egyptian temples, built over a span of more than 3000 years, may have been oriented with respect to the point where Spica sets over the horizon. Owing to the effects of precession, this point shifts slightly over the years, and it may well be that this was one way in which the Greek astronomer Hippachus discovered precession, around the year 150 BC. Spica's own proper motion is very slight, since the star is over 250 light-years away.

Spica lies within a dozen degrees of the celestial equator, and is prominent in the night sky for several months in every year. It lies in a rather barren region, which is why it was sometimes known in ancient times as Al Simak al Azel, the Defenceless or Unarmed One (in contrast to the other Simak, Arcturus, which was equipped with a lance).

Telescopically Spica appears as a single white or slightly bluish star, but in 1889 the German astronomer Hermann Vogel, working from the Potsdam Observatory, discovered that it is a spectroscopic binary. He found that the lines in the spectrum are periodically doubled, and this indicated that two components were involved.

With a binary, each star moves round the common centre of gravity of the system. If the orientation is suitable, this means that one component is approaching us while the other is receding (due allowance being made for the overall motion of the system, of course), and the lines will be Doppler shifted in opposite directions. This accounts for the doubling; in the case of Spica, the revolution

Equatorial Stars

Stars very close to the celestial equator are useful to the observer who wants to measure the angular diameter of his telescopic field. The procedure is as follows:

Put the star at one edge of the field, and allow it to drift centrally across. Measure the time taken for a full crossing, in minutes and seconds. Multiply by 15; the answer will give the diameter of the field in minutes and seconds of arc.

There are surprisingly few bright stars within one degree of the equator. Among them are:

Star	Magnitude	R.A.	Dec.
Delta Orionis	2.3	05h 32m 00.3s	-00° 17' 57"
Theta Aquilae	3.2	20h 11m 18.1s	-00° 49' 17"
Zeta Virginis	3.4	13h 34m 41.5s	-00° 35' 46"
Zeta Aquarii	3.6	22h 28m 49.9s	-00° 01' 13"
Delta Ceti	4.1	02h 39m 28.9s	+00° 19' 43"

As a matter of interest, the nearest first-magnitude star to the equator is Procyon (declination +5° 13'30").

Venus (centre) and Spica (bottom) below the Moon.
Photograph: John Dunlop, Auckland Observatory

Hermann Carl Vogel

Vogel, one of the great pioneers of stellar spectroscopy, was born in Leipzig in 1841. He assisted Zöllner in solar work, mainly in connection with prominences, and then became director of a private observatory in Bothkamp, where he specialized in studies of radial velocities. In 1874 he went to the Potsdam Observatory, and was Director from 1882. It was from here that he made the first detections of spectroscopic binaries, in 1889. In 1883 he had published the first catalogue of radial velocities. He died in 1907.

period is 4.014 days. With other pairs only one spectrum is visible, but the lines shift to and fro around their mean position, so that the period can be worked out.

The two Spicas are not equal; the primary has a diameter almost 11 times that of the Sun (9,500,000 miles or 15,300,000 km), and a

Gamma Virginis

Gamma Virginis, known by several alternative proper names (Arich, Porrima and Postvarta) is one of the most celebrated visual binaries in the sky. Each component is of magnitude 3.5; both are of type F0, and the luminosity is 7 times that of the Sun. The distance from us is 36 light-years.

The revolution period is 171.4 years; the real distance between the components ranges between 3 and 70 astronomical units, so that the orbit is decidedly eccentric. In 1836 they appeared so close together that with the telescopes of the time they were almost impossible to separate. They then widened, until Gamma became so wide and easy that almost any telescope would separate them. They are now closing up again; the present separation is 3 seconds of arc. The separation will be reduced to less than one second of arc by 2010, and for some years after that the star will again appear single except in giant telescopes.

mass 11 times that of the Sun, while the secondary seems to have 6 times the diameter of the Sun and 4 times the solar mass. Both are of spectral type B. The distance between them, centre to centre, is only about 11,000,000 miles (roughly 18,000,000 km), and this leads to a rather interesting situation. The orbits are of low eccentricity, and are inclined by about 24 degrees to the edgewise position as viewed from Earth. At each revolution a small part of the brighter component, which sends us 80 per cent of the total light of the system, is blocked out by the secondary star, and this causes a slight fall in

The "Bowl" of Virgo

Spica lies at the foot of the famous Y-form of Virgo. The "bowl" is formed to one side by Epsilon, Delta, Gamma, Eta and Beta Virginis, and on the other side by the second-magnitude Denebola (Beta Leonis). The "bowl" stars of Virgo are:

Star	Magnitude	Spectrum	Luminosity, Sun=1	Distance, lt-yrs	
Epsilon	2.83	G9	60	104	Vindemiatrix
Delta	3.38	M3	180	147	Minelauva
Gamma	2.75	F0+F0	7	36	Arich
Eta	3.89	A2	26	104	Zaniah
Beta	3.61	F8	120	147	Zavijava

Both Eta and Beta were ranked of the third magnitude in ancient times, but, as usual, the evidence in favour of real change is very slender. Delta is a fine red star whose colour is very strong when viewed with binoculars.

The "bowl" contains many members of the Virgo Cluster of galaxies, which includes many hundreds of members. Some of these are of special interest, such as the giant radio galaxy M. 87 (Virgo A). The average distance of the members of the Virgo cluster is 50,000,000 light-years.

Galaxies in the Virgo cluster.
200-inch photograph, Palomar

magnitude. To complicate matters, the primary is itself a pulsating variable. However, both effects combined produce only very slight fluctuations, amounting to only one-tenth of a magnitude. The changes cannot be detected with the naked eye, and need sensitive measuring equipment.

An inhabitant of a planet orbiting Spica would indeed have a remarkable view; two suns, one more brilliant than the other, describing a sort of waltz in the sky. True, Spica does not seem to be a promising candidate as a planetary centre, but one never knows.

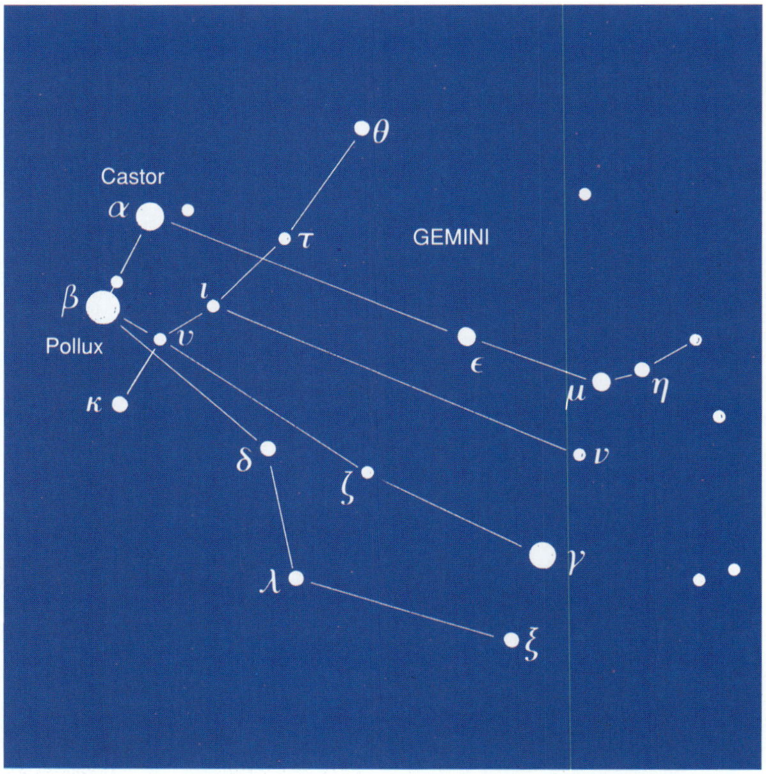

Diagram of Gemini, showing Castor and Pollux.

Gemini with slight trail. The bright object near the centre of the image is Saturn.
Photograph: Patrick Moore, 2-minute time exposure.

17. POLLUX

Beta Geminorum : 78 Geminorum

Right ascension:	07h 45m 18s.9
Declination:	+28° 01' 34"
Apparent magnitude:	1.14
Absolute magnitude:	0.2
Spectral type:	K0
Luminosity, Sun=1:	60
Distance, light-years:	36
Parallax:	0".092
Radial velocity:	+3 km/s
Proper motion (per year):	RA -0".047, dec. -0".05

Gemini is part of Orion's retinue, and is one of the brightest constellations in the Zodiac. Its two main stars are the Twins, Castor and Pollux; the rest of the constellation consists of chains of stars extending from the Twins in the general direction of Betelgeux

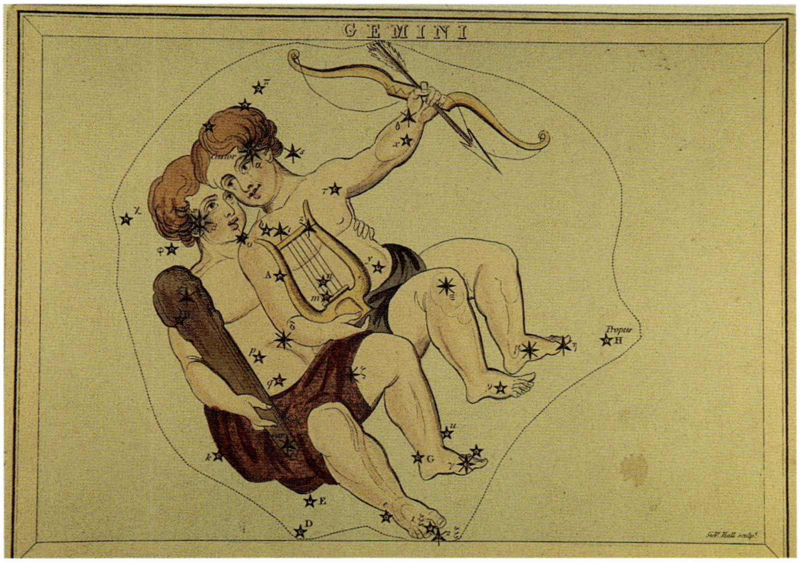

Old figure of Gemini.

in Orion. Gemini is crossed by the Milky Way, and the whole area is very rich.

Of the two leaders Pollux is the brighter by almost half a magnitude, though it is lettered Beta, and traditionally Castor was always regarded as the senior Twin. At the moment Pollux is officially classed as being of the first magnitude, while Castor is not, though in any case the dividing line is entirely arbitrary; less than two-tenths of a magnitude separates Regulus, the last of the élite, from Adhara in Canis Major, the first of the "also rans".

Curiously, ancient observers ranked the Twins as equal; Ptolemy around 150, Al-Sufi in 960, Ulugh Beigh in 1430 and Tycho Brahe in 1590 recorded them as being of the second magnitude. So too did Bayer in 1603, while in Flamsteed's list of 1700 it is Castor which is of the first magnitude and Pollux of the second. Only during the past two centuries has the superiority of Pollux been universally agreed. If any change has occurred (and to be candid, this is most improbable) it is more likely to have been in Pollux, as an orange late-type star, than in Castor, which is a multiple system.

There is a famous legend about Castor and Pollux. They were twin boys, with the same mother,- the immortal Leda, but different fathers. Pollux was immortal, while Castor was not. When the inevitable happened, and Castor was killed, Pollux was so distraught that he pleaded to be allowed to share his immortality with his brother. His wish was granted, and both boys were placed in the sky.

Angular Distance

In the sky Castor and Pollux are separated by 4½ degrees. The separation between Alpha and Beta Centauri is about the same. Bearing in mind that the angular diameter of the full moon is about half a degree, the following separations may be interesting:

Degrees (approx)	Stars
2	Altair to Gamma Aquilae
4½	Castor to Pollux; Alpha Centauri to Beta Centauri
5	Dubhe to Merak (in Ursa Major)
10	Betelgeux to Delta Orionis; Alpha Crucis to Beta Centauri
15	Alpha Centauri to Alpha Crucis
20	Betelgeux to Rigel
25	Vega to Deneb
35	Vega to Altair; Capella to Pollux
40	Capella to Betelgeux
45	Spica to Antares
50	Sirius to Castor; Vega to Polaris
60	Polaris to Pollux

Though the two stars appear side by side in the sky, there is absolutely no real connection between them. They are separated by at least ten light-years; Castor is the more remote. Perhaps the only comparable pair in the sky is that consisting of Alpha and Beta Centauri, but the Pointers to the Southern Cross are much brighter and more distinctive than the Twins.

Tycho Brahe's Star Catalogue

The best star catalogue of pre-telescopic times was compiled by one of the most colourful characters in the history of astronomy — Tycho Brahe.

Tycho was a Danish nobleman who became interested in astronomy at an early age — partly because of his observations of a brilliant supernova which flared up in Cassiopeia in the year 1572. He was proud and self-opinionated (during his student days he had part of his nose sliced off in a duel, and made himself a new one out of gold, silver and wax!) but of his brilliance there was no doubt at all. Supported by the King of Denmark, he established an observatory on the island of Hven, in the Baltic, and between 1576 and 1596 compiled the famous catalogue; he had no telescopes, but his work was amazingly accurate. He also made measurements of the positions of the planets, and it was these measurements which enabled his last assistant, Johannes Kepler, to prove that the Earth and the other planets move round the Sun, not in circular orbits, but in ellipses. Ironically, Tycho himself could never accept the new theories, and continued to believe that the Earth was the centre of the universe.

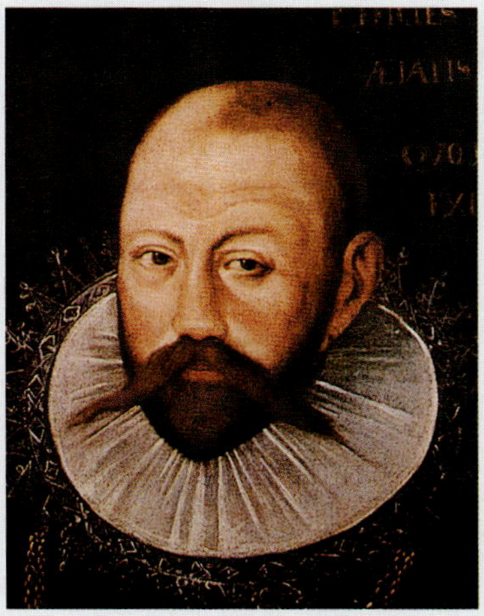

Tycho Brahe, last of the great pre-telescopic observers.

Tycho quarrelled with the Danish Court, and in 1596 left Hven to move to Prague as Imperial Mathematician to the Holy Roman Emperor. It was here that Kepler joined him, and when Tycho died suddenly, in 1601, Kepler obtained the priceless observations.

It was unfortunate that Tycho died before telescopes became available; he would have made such good use of them. After his death the observatory at Uraniborg, on Hven, was never used again; it fell into disrepair, but has recently been restored.

Stars in Gemini. Castor and Pollux are near the centre.
Photograph: S. Andrew

Pollux is a normal K-type star, with a diameter of around five times that of the Sun; over 4,000,000 miles (6,400,000,000 km). It seems to be constant in brightness, and its orange hue is very evident, particularly when compared with the pure whiteness of Castor. There are two faint companions, but both are over 200 seconds of arc away from Pollux itself and are unconnected with it. Indeed, both are slowly drifting away from Pollux and from each other.

There can be no problem in locating the Twins, which lie within 32 degrees of the celestial equator and are accessible from all inhabited countries for at least part of the year. Certainly they are worth finding, and it is at once clear that they are very definitely non-identical twins.

18. FOMALHAUT

Alpha Piscis Australis: 24 Piscis Australis

Right ascension:	22h 57m 38s.9
Declination:	-29° 37' 20"
Apparent magnitude:	1.16
Absolute magnitude:	2.0
Luminosity, Sun=1:	13
Distance, light-years:	22
Parallax:	0".141
Radial velocity:	+7 km/s
Proper motion (per year):	RA +0".026, dec. -0".16

Fomalhaut is the southernmost of the first-magnitude stars visible from anywhere in Europe or mainland United States. From these latitudes it is always so low down that British observers, in particular, tend to be surprised when they travel south and see Fomalhaut high in the sky; it is surprisingly bright. At declination almost -30 degrees it is circumpolar from anywhere south of latitude 60°S, and for much of the year it is well seen from countries such as New Zealand.

It lies in a somewhat isolated position in the sky, but it is easy to identify if the famous Square of Pegasus is on view. It lies in a direction indicated by two of the stars in the Square, Scheat and Markab; the line extended south from Markab crosses the large but dim constellation of Aquarius, the Water-bearer, which is included in the Zodiac. The only possible confusion is with Diphda or Beta Ceti, in the Whale, whose direction is indicated (more or less) by the other two stars of the Square, Alpheratz and Algenib. However, Diphda is a magnitude fainter than Fomalhaut.

This is a "watery" area of the sky map; as well as Aquarius, Cetus and Pisces (the Fishes) we have the Southern Fish, which may be called either Piscis Australis or Piscis Austrinus. Its leader is Fomalhaut, but otherwise it is very obscure, as there is no other star above the fourth magnitude, and neither is there any distinctive

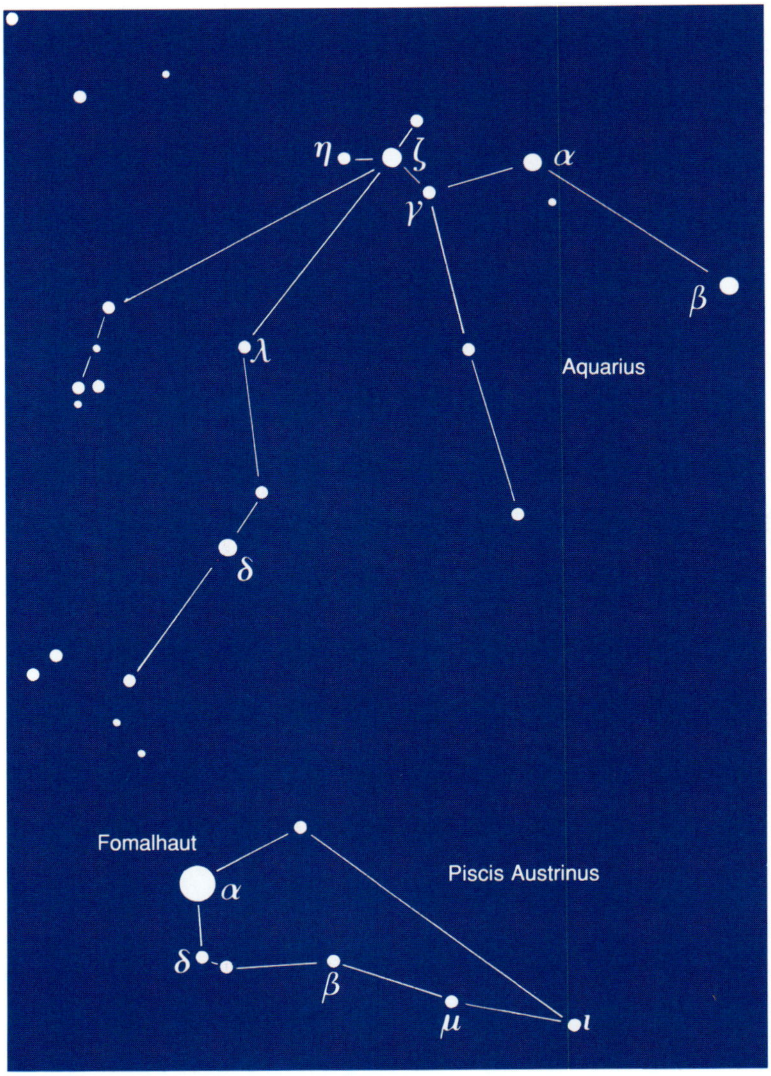

The Southern Fish.

Johann Mädler and the Centre of the Galaxy

Johann Heinrich von Mädler was born in Berlin in 1784. In 1822 he became a teacher, and gave private lessons in astronomy to Wilhelm Beer, a rich banker (and brother of the famous composer Meyerbeer). Using the 3.75-inch refractor in Beer's observatory, the two set out to study the Moon and planets. In 1837–8 they published a monumental work, *Der Mond,* giving a description of every named formation on the Moon; it was accompanied by a map which was a masterpiece of careful accurate observation, and which remained the standard for almost half a century.

In 1840 Mädler went to Dorpat, in Estonia (then part of Russia) as Director of the Observatory, and turned his attention to cosmology, in which he was notably less successful. In 1846 he claimed that the central star of the Galaxy was not Fomalhaut, but Alcyone in the Pleiades. The idea never met with much support, and was unsupported by any tangible evidence. It is, of course, completely wrong; the real centre lies beyond the star-clouds of Sagittarius, nearly 30,000 light-years away.

Mädler retired in 1865, and devoted his later years to research into the history of astronomy. He died in Hanover in March 1874.

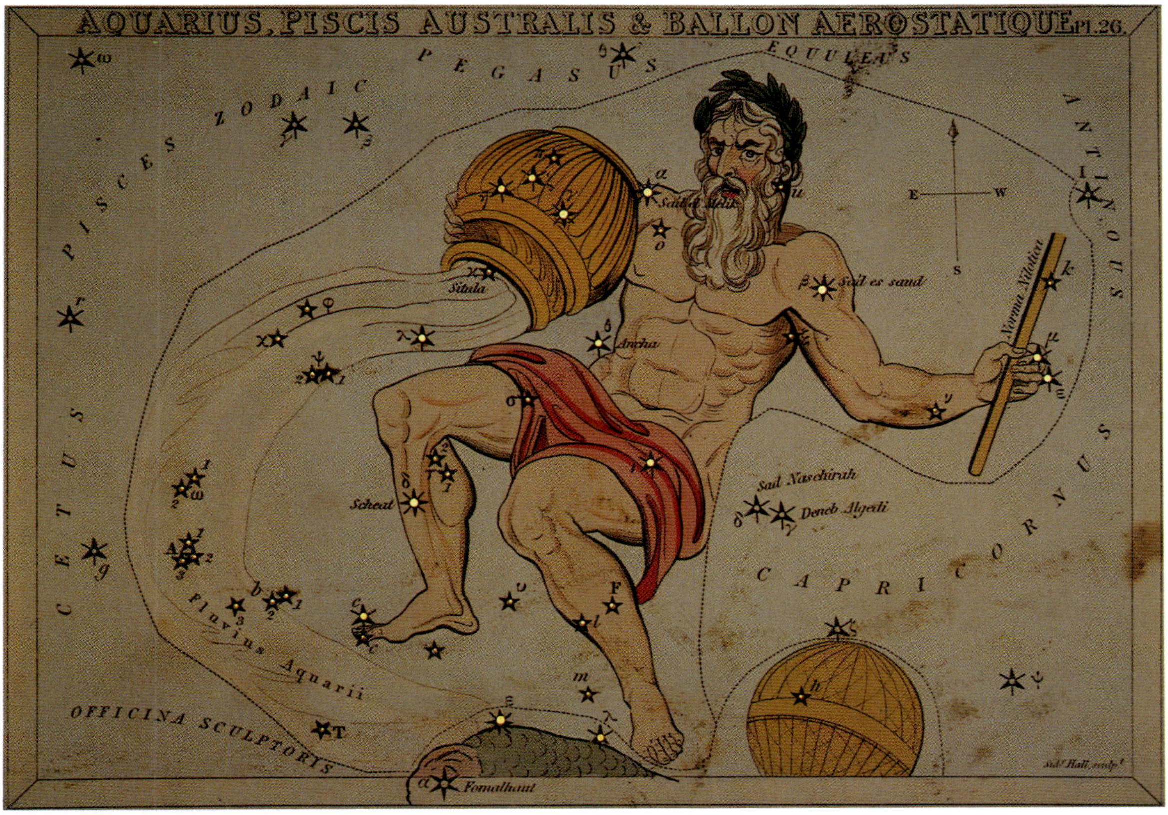

Old figures of the Southern Fish and Aquarius. Also shown is the now-rejected constellation of the Balloon.

pattern. It was one of Ptolemy's original 48 constellations, but there are no definite mythological legends attached to it apart from vague references to a Syrian water-god, Dagon, and the ferocious monster Typhon, who was said to be buried beneath Mount Etna.

The Greek astronomer Aratus, around 300 BC, referred to "one large and bright by both the Pourer's feet", clearly indicating Fomalhaut and Aquarius. To the Arabs, Fomalhaut was Al Difdi al Awwal, the First Frog (Diphda was Al Difdi al Thani, the Second Frog), though later it became Fum al Hut, the Fish's Mouth. In the Alphonsine Tables it was listed as "Fomalhant".

It should have been one of the four Royal Stars, together with Regulus, Aldebaran and Antares, since 4000 years ago it almost marked the winter solstice; but, as we have seen, its position well south of the celestial equator led to its replacement by the more accessible Altair. During the 19th century there was a curious suggestion that it might be the central star of the Galaxy. It is not easy to explain this idea; in fact Fomalhaut is one of our nearer neighbours, and is at least 25,000 light-years from the galactic centre.

Fomalhaut has two companions, both faint. At a separation of 30 seconds of arc there is a star of magnitude 14, but there is no real connection; the dim star merely lies in the same direction as seen from Earth. However the red dwarf GC 31978 (magnitude 6.5), two degrees south of Fomalhaut, is at about the same distance and has the same motion in space. Since the real distance between the two is almost a light-year it is hard to classify them as making up a binary system, but it has been suggested that the two may have had a common origin, so that originally both were members of a low-density open cluster.

Open clusters are common enough; the Pleiades and the Hyades in the northern sky, and the Jewel Box in the Southern Cross, are prime examples. Clusters of this sort may contain from a few tens of members to many hundreds; they are not tightly bound by gravitational forces, and are therefore subject to perturbations from non-associated stars, so that the cluster gradually disperses. This means that most open clusters are relatively young by cosmical standards, so that they have had no time to be scattered, though there are some

Ptolemy's Constellations

Ptolemy, last of the great astronomers of the ancient world, lived from about AD 120 to 180. He gave a list of 48 constellations, all of which are still in use, though in most cases with altered boundaries; others have been added since, notably in the far south of the sky, where they never rose from the latitude of Alexandria, where Ptolemy spent all his life.

Some of the "original 48" were large and important; others were small and dim — and Piscis Australis is one of the least conspicuous; it is redeemed only by the presence of Fomalhaut. The full Ptolemaic list is as follows:

Northern	
Ursa Minor	The Little Bear
Ursa Major	The Great Bear
Draco	The Dragon
Cepheus	Cepheus
Boötes	The Herdsman
Corona Borealis	The Northern Crown
Hercules	Hercules
Lyra	The Lyre
Cygnus	The Swan
Cassiopeia	Cassiopeia
Perseus	Perseus
Auriga	The Charioteer
Ophiuchus	The Serpent-bearer (originally Serpentarius)
Serpens	The Serpent
Sagitta	The Arrow
Aquila	The Eagle
Delphinus	The Dolphin
Equuleus	The Little Horse
Pegasus	Pegasus
Andromeda	Andromeda
Triangulum	The Triangle

exceptions, notably M.67 near Acubens in Cancer (the Crab), which lies in a sparse area almost 1500 light-years above the main plane of the Milky Way and is less subject to disruption; its age has been given as at least 4000 million years. It is quite logical to suppose that Fomalhaut and the red dwarf were formerly members of a cluster which has now been spread around so that it has lost its identity altogether, but there is no proof either way.*

Fomalhaut itself is a white A-type star. Some earlier reports described it as a red, but undoubtedly this was because these reports stem from the fact that to European observers the star is always so low that its light is reddened by passing through our atmosphere.

*Globular clusters, such as Omega Centauri and 47 Tucanae, are quite different; they may contain at least a million stars, and are compact symmetrical systems, capable of persisting almost indefinitely. Most of them lie round the edges of the Milky Way Galaxy. Other galaxies, too, contain associated globular clusters.

Zodiacal

Aries	The Ram
Taurus	The Bull
Gemini	The Twins
Cancer	The Crab
Leo	The Lion
Virgo	The Virgin
Libra	The Scales (originally Chelae Scorpii, the Scorpion's Claws)
Scorpius	The Scorpion
Sagittarius	The Archer
Capricornus	The Sea-goat
Aquarius	The Water-bearer
Pisces	The Fishes

Southern

Cetus	The Whale
Orion	Orion
Eridanus	The River
Lepus	The Hare
Canis Major	The Great Dog
Canis Minor	The Little Dog
Argo Navis	The Ship Argo
Hydra	The Water-snake
Crater	The Cup
Corvus	The Crow
Centaurus	The Centaur
Lupus	The Wolf
Ara	The Altar
Corona Australis	The Southern Crown
Piscis Australis	The Southern Fish

IRAS and ISO

IRAS, the Infra-Red Astronomical Satellite, made the observations which revealed clouds of cool material round Vega, Fomalhaut and other stars. It was built by a Dutch firm with help from the United States, and carried a Ritchey-Chrétien telescope of 22 inches aperture. This telescope was composed of 62 infra-red detectors cooled to -270ºC. by the use of liquid helium. It also carried various other instruments, including a spectrometer.

It was launched from Vandenberg Air Force Base in January 1983, and put into an orbit at an altitude of 550 miles (880 km). It operated for eleven months, and was a tremendous success. Over 2000 galaxies were charted; a full infra-red map of the sky was completed; infra-red material round stars; a ring of dust in the Solar System; graphite dust in interstellar space, and much else. It only ceased to operate when its supply of helium coolant ran out.

The latest infra-red satellite is ISO (the Infra-red Space Observatory), launched on 17 November 1995. It is moving in an eccentric orbit, taking it from 917 to 43,500 miles (1000 to 70,000 km) from the Earth; the orbital period is 24 hours. It carries an 0.6-metre infra-red telesceope, and is much more sensitive than IRAS.

Star trails taken from Arizona in 1994, where Fomalhaut rises to a respectable height above the horizon.
Photograph: Patrick Moore

In 1983 instruments on board the infra-red satellite IRAS showed that Fomalhaut, like Vega, is associated with a cloud of cool material which may well be planet-forming. Indeed, as a possible planetary centre Fomalhaut is rather more promising than Vega, because it is less luminous and less energetic; it is a very run-of-the-mill Main Sequence star. Of course there is no positive evidence in favour of a planetary family there, but we cannot rule it out, and there is at least a chance that on one of these hypothetical planets there is at this moment an intelligent astronomer looking across space toward a reasonably bright star — our Sun. With adequate radio equipment, he could even pick up broadcasts transmitted from Earth twenty-two years ago!

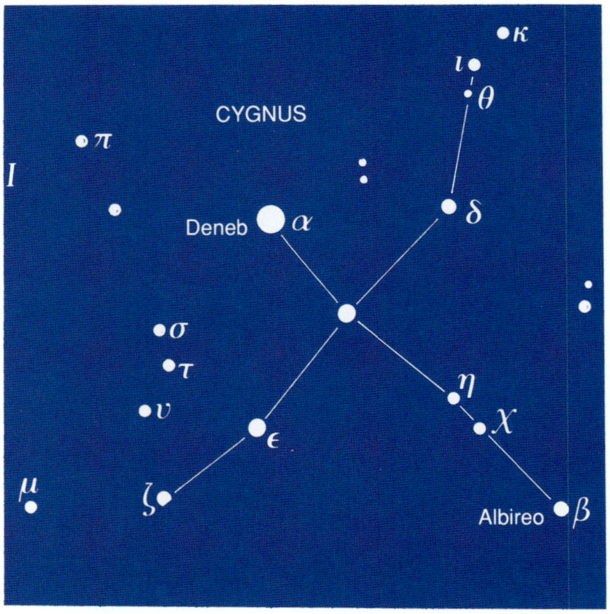

Cygnus (The Swan).

Old figure of Cygnus and its neighbours.

19. DENEB

Alpha Cygni : 50 Cygni

Right ascension:	20h 41m 25s.8
Declination:	+45° 16' 49"
Apparent magnitude:	1.25
Absolute magnitude:	-7.5
Spectral type:	A2
Luminosity, Sun=1:	70,000
Distance, light-years:	1800
Parallax:	0".007
Radial velocity:	-5 km/s
Proper motion (per year):	RA 0".000, dec. +0".01

Cygnus, the Swan, is one of the most striking of all the constellations. It is often nicknamed the Northern Cross, for obvious reasons; its five main stars are arranged in an X-pattern, much more cruciform than the kite-shaped Southern Cross. It is true that one of the five stars, Albireo or Beta Cygni, is fainter than the rest and further away from the centre, though to make up for this it is probably the loveliest coloured double in the sky.

Cygnus has always been associated with a bird; the Greeks called it Ορψιζ. The most celebrated legend concerns one of the many dubious adventures of Jupiter or Zeus, the king of the gods. He became attracted by Leda, wife of King Tyndaerus, and in order to be with her without causing undue attention he changed himself into a swan. The result was that Leda laid an egg, from which hatched the Twins, Castor and Pollux. It then was found that while Zeus really was the father of Pollux, Castor came from a totally different liaison — which is why Pollux was immortal, Castor not.

The name Deneb comes from the Arab Al Dhanab al Dajajah, the Hen's Tail. An alternative was Arided, from Al Ridf, the Hindmost, but this has long since become obsolete.

Deneb lies 45°N of the celestial equator; it is therefore circumpolar from Britain, but from parts of Australia and South Africa it is always low down, and from Invercargill at the southernmost tip of New Zealand it barely rises at all. It makes up a large triangle with Vega and Altair, but it appears fainter than either of them because it is so far away. Its distance has been given as 1800 light-years, so that it is the remotest of the first-magnitude stars; we see it today as it used to be at the time when Britain was still occupied by the Romans. Up to and including the time of Flamsteed (circa 1700) it was generally ranked of the second magnitude, but there is no suggestion of change; it all depends where the official list of "first magnitude" stars ends, and this is quite arbitrary. Some of the early catalogues cut it off at Fomalhaut.

Despite this vast distance, Deneb is still the 19th brightest star in the sky. It is a true cosmic supergiant; the luminosity is of the order of 70,000 times that of the Sun, in which case it outranks Rigel, and in our list is surpassed only by Canopus (whose real power is much less certain; as we have seen, 200,000 times the luminosity of the

The Cross of Cygnus

The five stars of the Northern Cross are quite unrelated to each other, so that once more we are dealing with nothing more significant than a line of sight effect. Deneb is much the most remote and most luminous. Details are as follows:

Star		Apparent magnitude	Absolute magnitude	Spectrum	Distance, lt-yrs
Alpha	Deneb	1.25	-7.5	A2	1800
Gamma	Sadr	2.20	-4.6	F8	750
Epsilon	Gienah	2.46	0.2	K0	81
Delta		2.87	-0.6	A0	160
Beta	Albireo	3.08	-2.3	K5	390

Albireo is a superb double, with a golden-yellow primary and a vivid blue companion of magnitude 5.1. The separation is over 34 seconds of arc, so that the pair can be split with almost any telescope or even good binoculars. Unquestionably the two components are associated, but the real distance between them is of the order of 4400 astronomical units (over 400,000 million miles), so that the revolution period must be immensely long. Delta has a 6th-magnitude companion at a separation of 2.4 seconds of arc, position angle 225º; it is a binary, with a computed period of 828 years, but is not too easy to split with any telescope below 3 inches (7.5 cm) aperture. Between Sadr and Albireo is a star of magnitude 3.89, Eta Cygni, close beside which is the red long-period variable Chi Cygni.

Cygnus is worth sweeping with binoculars or a low-power telescope; there are star-clouds, dark rifts in the Milky Way, and several clusters and nebulae.

A telescopic view of Albireo which is a superb double.
Photograph: Paul Doherty.

Sun may be an over-estimate). Deneb has an A-type spectrum, so that it is pure white. Like many very powerful stars it is slightly variable, but the range is too slight to be detected except with sensitive measuring instruments. Its mass may be as much as 25 times that of the Sun, so that by stellar standards it is very much of a heavyweight. Its life-span is bound to be relatively short, and it will end its brilliant career by exploding as a supernova.

Deneb lies in a rich part of the Milky Way. Near it in the sky we find one of the most interesting nebulae in the northern hemisphere. It is known officially as NGC 7000 (the 7000th object in the New General Catalogue drawn up in the 1880s by J.F.R. Dreyer; it is not included in Messier's catalogue but is C20 in the Caldwell catalogue), and is always known by its nickname of the North America Nebula. The bright nebulosity really does give the impression of the shape of the North American continent, with the Gulf of Mexico marked

Opposite *The North America Nebula in Cygnus, near Deneb.* 48-inch Schmidt photograph, Palomar.

by a dark mass. The Nebula is visible with the naked eye, though it looks like nothing more than a slightly brighter part of the Milky Way; binoculars bring it out well, and with a powerful pair, say 20x70, the shape is easy to make out. Obviously, it is a favourite target for astrophotographers.

Bright nebulae are of two types: reflection, and emission. A reflection nebula shines only by being lit up by stars in or near it. With an emission nebula, the stellar radiation causes the nebulosity to emit a certain amount of light on its own account, and this is the case with the North America Nebula. The real distance from Deneb is around 70 light-years, but such is the immense brilliance of Deneb

Highly Luminous Stars

Deneb is one of the most luminous of the naked-eye stars. Among other "searchlights" which appear in our skies as above the fourth magnitude are:

Star		Apparent magnitude	Absolute magnitude	Spectrum	Distance, lt-yrs
Alpha Carinae	Canopus	-0.72	-8.5	F0	1200
S Carinae		4.68	-8.5	F0	11800
Epsilon Aurigae	Almaaz	2.99v	-8.5	F0	4600
b Velorum		3.54	-8.4	F2	6200
Delta Canis Majoris	Wezea	1.86	-8.0	F8	3060
Rho Cassiopeiae		4.8v	-8.0	F8	5400
Beta Doradûs		3.7v	-8.0	F9	7500
Nu Cephei		4.29	-7.5	A2	3900
Alpha Cygni	Deneb	1.25	-7.5	A2	1800
H Camelopardalis		4.21	-7.1	B9	4000
Mu Sagittarii	Polis	3.86	-7.1	B8	3900
Beta Orionis	Rigel	0.12	-7.1	B8	910
Eta Canis Majoris	Aludra	2.44	-7.0	B5	2500
Mu Cephei		4.8v	-7.0	M2	1560
Kappa Orionis	Saiph	2.06	-6.9	B0.5	2100
Omicron^2 Canis Majoris		3.86	-6.8	K3	2800
Chi^2 Orionis		4.63	-6.8	B2	3260
Zeta^1 Scorpii		4.73	-6.7	B1.5	2500
Kappa Cassiopeiae		4.16	-6.6	B1	3000
Alpha Camelopardalis		4.29	-6.2	O9.5	2800
Epsilon Orionis	Alnilam	1.70	-6.2	B0	1200
Delta Orionis	Mintaka	2.23v	-6.1	O9.5	2350
Phi Velorum		3.54	-6.0	B5	2500
Omicron^1 Canis Majoris		3.86	-6.0	K3	2800
Tau Canis Majoris		4.39	-6.0	O9	3600
Iota Orionis	Hatysa	2.76	-6.0	O9	1860
Zeta Orionis	Alnitak	1.77	-5.9	O9.5	1100

A star of absolute magnitude -8.5 has a brightness approximately 200,000 times that of the Sun; absolute magnitude -7, 52,500 Suns; absolute magnitude -6, 20,900 Suns.

that it seems to play a major rôle in illuminating and exciting the nebular material, though no doubt other stars are also contributors. The distance of the Nebula is about the same as that of Deneb; its real diameter is almost 50 light-years, so that if it were centred on the Sun it would contain stars such as Sirius, Procyon and Fomalhaut. Of course, the material in it is immensely rarefied. Hydrogen is the main constituent, together with a great deal of "dust" which is highly effective at blocking out the light of objects beyond.

From Deneb, our Sun would appear as a dim star of about the 14th magnitude. It is most improbable that there are any attendant planets; Deneb is far too young and far too energetic to be a promising candidate, though, as always, we cannot be sure.

Binoculars and the Sky

Binoculars are very useful in astronomy, and in fact a good pair of binoculars is far better than a very small telescope. They are classified according to their aperture (always given in millimetres) and magnification; thus a pair of 7x20 binoculars gives a magnification of 7, with each object-glass 20mm in diameter. In fact, a pair of binoculars is made up of two small refractors joined, and working together.

The main advantage is ease of use. If only one pair is to be bought, it is wise not to exceed a magnification of about 12; more powerful binoculars are heavy to hold, and need some sort of mounting (a converted camera tripod will do well, though neck mounts can also be bought or made).

The main disadvantage is lack of sheer magnification. Binoculars will not, for example, show the rings of Saturn, but they will provide endless enjoyment; the craters of the Moon, the phases of Venus, the Galilean satellites of Jupiter, and many double stars, coloured stars, clusters and nebulae. The one thing that must never be done, of course, is to look at the Sun through any telescope or pair of binoculars, even with the addition of a dark filter.

20. BETA CRUCIS

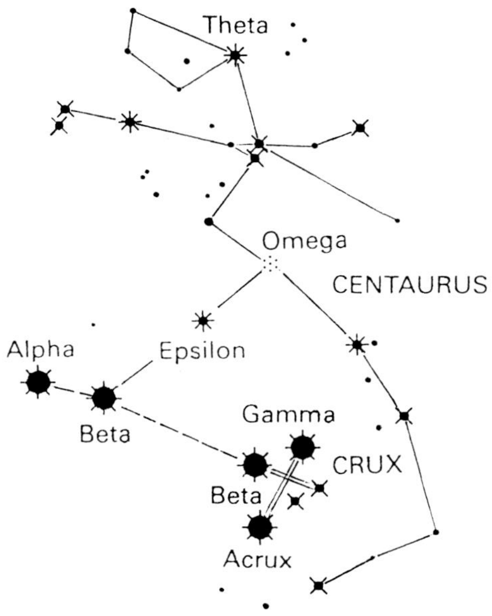

Crux and Centaurus.

Right ascension:	12h 47m 43s.2
Declination:	-59° 41' 19"
Apparent magnitude:	1.25
Absolute magnitude:	-5.0
Spectral type:	B0
Luminosity, Sun = 1:	8200
Distance, light-years:	425
Parallax:	0".007
Radial velocity:	+20 km/s
Proper motion:	RA -0".005, dec. -0".02

The second star of the Southern Cross has never had an accepted proper name. Here and there it has been called "Mimosa", but most people simply refer to it as Beta Crucis.

It is a hot, bluish-white B-type star, more powerful than the two components of Alpha Crucis combined. It is also further away. In some earlier lists it brought up the rear of the "first-magnitude table", but this distinction now belongs to Regulus in Leo, which is slightly the fainter of the two. There is a less than 4/10 of a magnitude between Beta Crucis and the red star Gamma Crucis, but Gamma has always been ranked as of the second magnitude; as we have noted, the cutting-off point is quite arbitrary.

Beta Crucis is slightly variable, though with a very small range of less than 0.1 of a magnitude; the period is 0.2365 days. It is pulsating, and changing its output slightly as it does so. Stars of this kind are known as Beta Cephei or Beta Canis Majoris variables, after the two prototypes; all are giants or sub-giants, mainly of type B, with luminosities several thousand times that of the Sun. The fluctuations are much too tiny to be detectable with the naked eye.

Perhaps Beta Crucis is notable mainly because of its surroundings. Close beside it in the sky (though not in reality) are two particularly famous objects: the Jewel Box cluster, and the Coal Sack.

The Coal Sack is the best example of a dark nebula. It looks like a

John Herschel at the Cape

Sir William Herschel, discoverer of the planet Uranus, in 1781, carried out a full survey of the northern hemisphere of the sky, using telescopes which he made himself; he discovered thousands of double stars, clusters and nebulae, and also drew up the first reasonably good picture of the shape of the Galaxy, though admittedly he erred in placing the Sun near the centre of the system. However, he spent all his active career in England, and so could not see the stars of the far south.

William died in 1822. His son, John (later Sir John) decided to complete the survey by extending it to the southern hemisphere, and so in January 1834 he arrived at the Cape of Good Hope, bringing with him an excellent telescope, of 20 feet focal length, made by him and his father.

He set it up at Feldhausen, outside Cape Town. By the time he left for home, in March 1838, the survey had been completed. He observed 1707 nebulae, of which 1268 were new; he listed 2102 binary pairs; he made magnificent drawings of the Large Magellanic Cloud and the unique nebula round the erratic variable star Eta Carinae (then known as Eta Argûs), and followed the fluctuations of Eta itself, which was then one of the most brilliant stars in the sky and at times outshone even Canopus. He estimated the luminosities of some stars, making Alpha Centauri 4 times as powerful as the Sun, Vega 40 times and Arcturus 200 times, which was a remarkably good attempt. Solar activity was monitored, and a long series of drawings and measurements was made of Halley's Comet, which was on view for the first time since 1759, and was not to return again until 1910.

John Herschel.

He made one amusing entry in his diary: "We had a fine view of the Comet on his return from the Sun. I followed him until about the 20th day of May — and should have kept him in view for longer, but for a naughty nebula not down in my Catalogue which came so near to his place and looked so like him that he fairly led me off the scent. To say the truth, I am glad he is gone." Herschel was, in fact, the last to see the comet at its 1835 return.

The Cape trip was an unqualified success, and it is fair to say that John Herschel was the real founder of southern-hemisphere stellar astronomy. The Jewel Box and the Coal Sack were only two of the many objects which came under his scrutiny. Today the Feldhausen site is occupied by the Grove Primary School, but there is a majestic obelisk marking the exact position where John Herschel's telescope once stood.

A model of the telescope taken by John Herschel to the Cape.

Supernova 1987a, with the Tarantula Nebula and the Large Cloud of Magellan.

"hole in the heavens", to use William Herschel's graphic term, but it is in fact a huge mass of obscuring material which effectively blocks out the light of objects behind it; the degree of obscuration ranges from 50 per cent in some parts of the cloud to over 90 per cent in others. With the naked eye, only one dim star can be seen in front of it. Against a dark sky, and in the absence of moonlight, it is very evident, and it has been known since very early times.

There are many legends about it. The Australian Aborigines said that it was "the embodiment of evil in the shape of an Emu, who lies in wait at the foot of a tree, represented by the stars of the Cross, for a possum driven by his persecutions to take refuge among its branches". The Peruvians had a kinder view of it; to them it was a celestial Doe suckling its fawn. It was recorded in 1499 by a traveller named Vincente Pizon, and Amerigo Vespucci (he who gave his name to America) called it "Canipo fosco", the Dark Canopus. It was also nicknamed the Black Magellanic Cloud. One writer gave a description of it which was vivid, if inaccurate: "An inky spot, an opening into the awful solitude of unoccupied space".

The Coal Sack has a longest diameter of 7 degrees, and covers a total area of just over 26 square degrees; the real diameter is almost 70 light-years. It is 550 light-years away, beyond any of the main members of the Cross; in a way it is rather surprising that there are not more foreground stars.

A bright nebula, such as M.42 in Orion's Sword, shines because it is being lit up by stars in or near it; with M.42, the illuminating stars are those of the Trapezium, Theta Orionis. (In fact, M.42 is an emission nebula; the radiation from the very hot stars makes the nebulosity give off a certain amount of light on its own account.) It

M.42, the Great Nebula in Orion, photographed with the 200-inch reflector at Palomar.

so happens that with the Coal Sack there are no suitable stars available, so that the nebulosity remains dark. But this might not be the case if we were looking at it from a different vantage point. Who knows what may lie on the far side of the Coal Sack, forever hidden from our eyes? There may be stars there which make the nebulosity shine, so that an observer would see it as gloriously bright. It is equally possible that there are no illuminating stars on the far side of M.42, so that if we lay on the other side of it we might see it looking as black as the Coal Sack does to us.

There are plenty of dark nebulae in the sky, and there are dark

Barnard's Dark Nebulae

One of the greatest American astronomers of modern times, Edward Emerson Barnard, was born in Tennessee in 1857. He became a photographic apprentice, but showed an early aptitude for astronomy, and specialized in searching for comets; he made several discoveries.

He joined the staff of the Lick Observatory in 1888, and while there discovered Amalthea, the fifth satellite of Jupiter. In 1897 he moved to the Yerkes Observatory, and specialized in stellar studies; he discovered the fast-moving star in Ophiuchus which we still call Barnard's Star (our nearest neighbour apart from the members of the Alpha Centauri group), and began a systematic photographic survey of the sky. During the course of this survey he discovered many dark nebulae, and in 1919 published the first catalogue of them; he correctly assumed that they were due to obscuring matter. He died in 1923.

rifts in the Milky Way, but certainly there is nothing quite so eye-catching as the Coal Sack. One can well understand why it acquired its nickname.

Also in Crux, and also close to Beta, is the Jewel Box cluster round Kappa Crucis, officially known as NGC 4755 (position: RA 12h 54m, declination -60° 20'). It is C 94 in the Caldwell Catalogue. It was so called by Sir John Herschel, who observed it during the 1830s when he was working at the Cape. It has been said that the cluster contains stars of many different colours, but this is an exaggeration; most of the leaders are hot, bluish-white and very luminous, though there is one prominent red supergiant which stands out at once — and is probably just as powerful as Betelgeux. The cluster is visible with good binoculars, or with almost any telescope, though it is true that a somewhat larger telescope is needed to show the cluster in its

The Southern Cross.
Photograph: Jack McBain, Rhodesia, 1958.

full glory. Three of its main stars make up a triangle, inside which lies the red supergiant.

The Jewel Box is about 7700 light-years away, and at the moment it is approaching us at a rate of 11 miles (18 km) per second, though needless to say this will not continue indefinitely, and there is no fear of an eventual collision on the line. The central region is around 25 light-years across; the cluster is young, with a probable age of only a few million years, which is why most of its leaders have not had enough time to evolve off the Main Sequence and move into the giant branch of the H-R Diagram.

It is sheer coincidence that the Jewel Box lies at the edge of the Coal Sack, because there is absolutely no connection between them. Yet when taken together, they provide a spectacle which certainly ranks as one of the show-pieces of the sky.

21. REGULUS

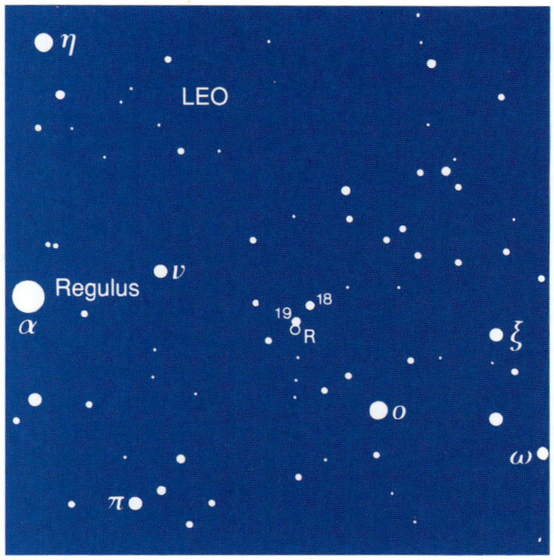

R Leonis.

Alpha Leonis: 32 Leonis

Right ascension:	10h 08m 22s.2
Declination:	+11° 58' 02"
Apparent magnitude:	1.38
Absolute magnitude:	-0.6
Spectral type:	B7
Luminosity, Sun=1:	130
Distance, light-years:	85
Parallax:	0".048
Radial velocity:	+4 km/s
Proper motion (per year):	RA -0".017, dec. 0".00

Leo represents the Nemaean Lion, a ferocious beast killed by Hercules as one of the celebrated Twelve Labours. In the sky, however, it must be said that the hero is much less imposing than his victim. Hercules is a large but decidedly dim constellation, with no star much above the third magnitude; Leo, in the Zodiacal band, is bright and distinctive, with one star (Regulus) of the first magnitude and another four above the third.

Opposite ***Opposite*** *An artist's impression of a Leonid Meteor storm.*
Paul Doherty.

There should be little trouble in locating Leo, which is at its best during evenings around March and April. If Ursa Major is on view, simply use the Pointers, Merak and Dubhe, "the wrong way" — that is to say, away from Polaris; if sufficiently prolonged, the line will indicate the curved line of stars marking what is termed the Sickle of Leo. If Ursa Major is unavailable, it is worth noting that Regulus forms a very large triangle with Spica and Procyon. The Sickle is prominent, and somewhat resembles the mirror image of a question-mark; the rest of the constellation is marked by three brightish stars forming a triangle. One member of the triangle, Denebola or Beta Leonis, was ranked as of the first magnitude by some old observers, though it is now slightly below the second.

Leo is a very ancient constellation. Between 4000 and 2000 BC it contained the solstice, and the Egyptian king Necepsos taught that

Old figure of Leo. Leo Minor, the Little Lion, is also shown.

at the time of the Creation the Sun rose in Leo, so that it was known as "Domicilis Solis", the emblem of fire and heat (both it and Sirius were commonly linked with the high temperatures and sudden storms of summer). Pliny recorded that Leo was worshipped because the entry of the Sun into the constellation coincided with the annual flooding of the Nile, upon which the whole Egyptian economy depended. In China it was the Yellow Dragon, but in general it was associated with a lion; even in the famous Dendereh Zodiac of Egypt it was depicted as a lion standing upon a serpent.

Regulus itself may be the faintest of the accepted first-magnitude stars, but it has always been regarded as important; it was one of the Royal Stars, and the name is derived from Rex. In India it was

Magha, the Mighty; in Persia it was Miyau, the Centre; to the Romans it was Cor Leonis, the Lion's Heart.

It lies less than a dozen degrees from the celestial equator, so that it can be seen from all inhabited parts of the world. It is also one of the few really bright stars which can be occulted by the Moon. Very occasionally it can be occulted by a planet, and this happened on 7 July 1959, when it passed behind Venus. At the time this was regarded as an important event, because before the actual occultation the light from Regulus would have to pass through the shell of atmosphere surrounding Venus, about which our knowledge was decidedly meagre. The phenomenon was fascinating (I observed it with a 12-inch reflector, and saw Regulus flicker and fade for some seconds before disappearing). A great deal of useful information was obtained, and it was not until some years afterwards that the space missions finally told us what Venus is really like.

In itself Regulus is ordinary enough; it is a normal B-type star, pure white in colour, with a diameter about five times that of the Sun and a surface temperature of 12,000°C. It has a 7.7-magnitude companion at a separation of 177 seconds of arc, and since it and Regulus share the same motion in space they are presumably associated; the companion is a close and difficult binary whose colour has been variously described as indigo, golden and orange-red, even though most observers will certainly call it white.

Close to Regulus in the sky is the Mira-type variable R Leonis, which can attain naked-eye visibility at maximum and never sinks

The Sickle of Leo

The Sickle is made up of seven stars, of which Regulus is the brightest. They are:

Star		Apparent magnitude	Absolute magnitude	Spectrum	Distance, lt-yrs
Alpha	Regulus	1.38	-0.6	B7	85
Gamma	Algieba	1.99	0.2	K0+G7	90
Epsilon	Asad Australis	2.98	-2.0	G0	310
Zeta	Adhafera	3.44	0.6	FO	117
Eta		3.52	-5.2	A0	560
Mu	Rassalas	3.88	-0.1	K2	200
Lambda	Alterf	4.31	-0.3	K5	260

Eta, which looks relatively insignificant, is much the most luminous of these stars — over 9000 times the power of the Sun. Gamma (Algieba) is a fine double; the components are of magnitudes 2.2 and 3.5, and the separation is 4.3 seconds of arc, so that a small telescope will split it. It is a binary, with a period of 619 years. The primary is of a yellow hue; the companion has been described as greenish (by contrast), though most observers will see it as white. The separation is very gradually increasing.

Trails of Leonid Meteors, 1966. This was the great Leonid "storm" which may be repeated in 1999.

The Leonid Meteors

Within the Sickle, only 2 degrees away from Algieba, lies the radiant of one of the most interesting of all meteor showers: the Leonids.

Meteors are cometary debris. As a comet moves along it leaves a "dusty" trail behind it; when the Earth passes through such a trail it collects small particles, usually smaller than grains of sand, which dash into the upper air and burn away, producing the luminous streaks which we call shooting-stars. (Note that there is no connection between a shooting-star and a meteorite, which may land intact and produce a crater; meteorites come from the asteroid belt.) There are many annual showers, most of them associated with known parent comets; thus the October Orionids are associated with Halley's Comet.

The parent comet of the November Leonids is Tempel-Tuttle, named after the two independent discoverers in 1866. The period of the comet is 33 years. The Earth encounters the trail each November, but major meteor "storms" are seen only when we pass through the densest part of the swarm,. that is to say, when the comet returns to perihelion. This happened in 1799, 1833 and 1866. There were no major showers in 1899 and 1933, because the orbit of the swarm had been perturbed by the planets, but the Leonids produced a magnificent display for a few hours on 17 November 1966. The next return of the comet is due in 1999, and another major 'storm' is likely. When the Leonids are at their best, it has been said that meteors seem to rain down like snowflakes.

Leo. The bright object in the Sickle of Leo, above Regulus, is the planet Mars.
Photograph: Christopher Doherty, 1995.

much below magnitude 11; it remains within the range of small telescopes, and there are useful comparison stars to hand. It is also convenient inasmuch as its period is 312 days, so that it comes to maximum between one and two months earlier each year. Other Mira stars are not so obliging. Thus R Serpentis, which also can reach naked-eye visibility, peaks only 9 days earlier each year, so that there are periods when successive maxima are more or less unobservable because they occur when the star is above the horizon only during the hours of daylight. Another instance of this sort of behaviour is provided by U Orionis, near Betelgeux, which comes to maximum about a week later each year. It would be irritating to find a Mira star with a period of exactly one year — though as we have seen, these variables are not constant in either period or amplitude from one cycle to another, and the periods may fluctuate by several days either way.

Also near Regulus in the sky are two galaxies, Leo I and Leo II, which are so faint that they are beyond the range of most telescopes

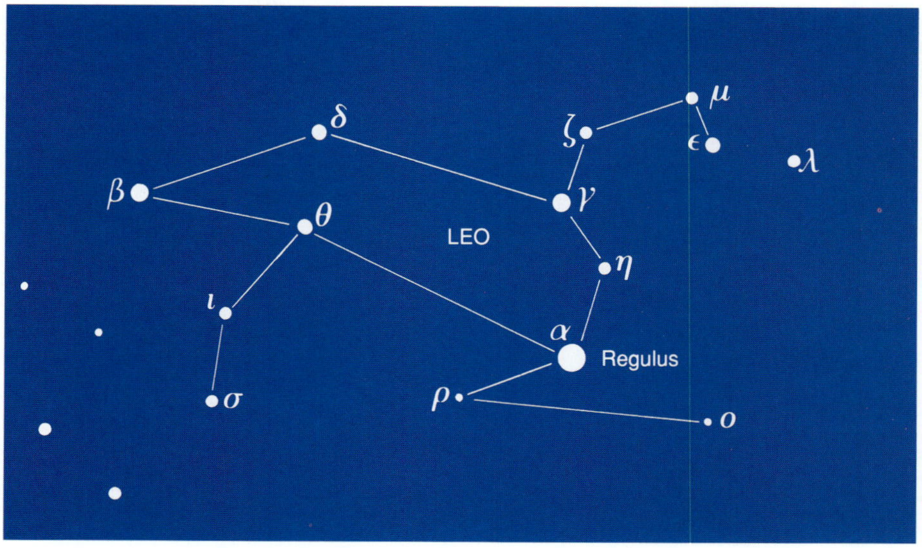

Leo (The Lion).

Nearby Dwarf Galaxies

Galaxies tend to occur in groups or clusters. Our Galaxy is a member of the Local Group; this is made up of three relatively large systems (our Galaxy, the Andromeda Spiral and the Triangulum Spiral), and more than two dozen much smaller members. Most of these are dwarfs, not easy to identify because they are faint and sparse. Several lie within a million light-years of us:

Galaxy	Type	Absolute magnitude	Distance, lt-yrs	Diameter, lt-yrs
Our Galaxy	Spiral	-20.5	-	100,000
Sagittarius dwarf	Elliptical	-	80,000	10,000
Large Magellanic Cloud	Barred spiral	-18.5	169,000	30,000
Small Magellanic Cloud	Barred spiral	-16.8	190,000	16,000
Ursa Minor dwarf	Elliptical	-8.8	250,000	2,000
Draco dwarf	Elliptical	-8.6	250,000	3,000
Sculptor dwarf	Elliptical	-11.7	280,000	5,000
Fornax dwarf	Elliptical	-13.6	420,000	7,000
Carina dwarf	Elliptical	-	550,000	-
Leo I	Elliptical	-11.0	750,000	2,000
Leo II	Elliptical	-9.4	750,000	3,000

The Andromeda Spiral lies at 2,200,000 light-years, and the Triangular Spiral at 2,900,000 light-years. The closest of all galaxies is the Sagittarius dwarf, identified in 1994; it is 30,000 light-years from the centre of the Milky Way galaxy, but on the far side, so that its distance from us is 80,000 light-years. It is being disrupted by the gravitational pull of our Galaxy, and will eventually lose its identity.

The two Magellanic Clouds are prominent naked-eye objects. At first glance they seem irregular in shape, but in each case there are traces of a barred spiral structure. They lie in the far south of the sky, and are inaccessible from any part of Europe.

and are also very difficult to photograph, because they are so sparse and so scattered. They belong to the Local Group, and may be regarded as satellites of our Galaxy; at any rate, they are gravitationally bound to it even though they are well over 7,000,000 light-years away. There are nearer and brighter galaxies in Leo, including several Messier objects, so that all in all Leo has a great deal to offer, quite apart from the fact that it is graced by the presence of a Royal Star.

MAGNITUDES 1.5 TO 2

Star		Apparent magnitude	Luminosity, Sun=1	Spectrum	Distance, lt-yrs
22. Epsilon Canis Majoris	Adhara	1.50	5000	B2	490
23. Alpha Geminorum	Castor	1.58	45	A0	46
24. Gamma Crucis		1.63	160	M3	88
24. Lambda Scorpii	Shaula	1.63	1300	B2	275
26. Gamma Orionis	Bellatrix	1.64	2200	B2	360
27. Beta Tauri	Al Nath	1.65	400	B7	130
28. Beta Carinae	Miaplacidus	1.68	130	A0	85
29. Epsilon Orionis	Alnilam	1.70	23000	B2	1200
30. Alpha Gruis	Alnair	1.74	230	B5	69
31. Zeta Orionis	Alnitak	1.77	19000	O9.5	1100
31. Epsilon Ursae Majoris	Alioth	1.77	60	A0	62
33. Gamma Velorum	Regor	1.78	3800	WC7	520
34. Alpha Ursae Majoris	Dubhe	1.79	60	K0	75
35. Alpha Persei	Mirphak	1.80	6000	F5	620
36. Epsilon Sagittarii	Kaus Australis	1.85	110	B9	85
37. Delta Canis Majoris	Wezea	1.86	130000	F8	3000
37. Epsilon Carinae	Avior	1.86	600	K0	200
37. Eta Ursae Majoris	Alkaid	1.86	450	B3	108
40. Theta Scorpii	Sargas	1.87	14000	G0	900
41. Beta Aurigae	Menkarlina	1.90	50	A2	46
42. Alpha Trianguli Aust.	Atria	1.92	96	K2	55
43. Gamma Geminorum	Alhena	1.93	82	A0	85
44. Alpha Pavonis		1.94	700	B3	230
45. Delta Velorum	Koo She	1.96	50	A0	68
46. Alpha Hydrae	Alphard	1.98	105	K3	85
46. Beta Canis Majoris	Mirzam	1.98v	7200	B1	710
48. Gamma Leonis	Algieba	1.99	60	K0+G7	90
48. Alpha Ursae Minoris	Polaris	1.99	6000	F8	680
50. Alpha Arietis	Hamal	2.00	96	K2	85

Excluding the unique Eta Carinae, there are two variable stars which may at maximum exceed the second magnitude:

Omicron Ceti	Mira	max 1.2?	140v	M	95
Gamma Cassiopeiae		max 1.6	6000v	B0	780

22. FROM REGULUS TO POLARIS

We have discussed the 21 brightest stars, those which are officially ranked as being of the first magnitude. Opposite we list those which follow on, from magnitude 1.5 to 2.0.

Adhara (Epsilon Canis Majoris) is one of three very hot, very luminous stars in Canis Major; they form a triangle well south of Sirius. The others are Delta (Wezea) and Beta (Mirzam). Adhara is an important star, because it has a great effect upon what is known as the Local Cloud, a huge, rarefied mass of hydrogen 50 light-years across, in which the Sun lies.

The mass of the Local Cloud is measured by the fact that it causes absorption lines in the spectra of stars beyond it. However, ionized hydrogen cannot produce these, and so we want to find out how much of the hydrogen is ionized and how much is neutral. At a wavelength of 600 Ångströms, Adhara is the most powerful emitter in the sky, so far as we know; at this wavelength it is fifteen times more brilliant than any other star, and it is responsible for most of the ionization in the Local Cloud. Indeed, it has probably ionized 15 to 19 per cent of all the hydrogen atoms in the Cloud. Originally it was thought that space would be more or less opaque to extreme ultra-violet radiation, but the EUV (Extreme Ultra-Violet Explorer) satellite showed that this is not so; in some directions there is little absorption of EUV, and this is particularly the case in the direction of Canis Major, because of the presence of Adhara.

Castor (the senior though fainter of the Twins) appears as a single white star to the naked eye, but it is in fact a complex system. It is a fine binary; the components are of magnitudes 1.9 and 2.9, with a period of 420 years. The separation is 2.5 seconds of arc. Each component is a spectroscopic binary, and there is a third component, about 100 thousand million miles (160 thousand million km) from the main pair which is itself a spectroscopic binary. The faint pair is known as Castor C, but also has a variable star designation — YY Geminorum — as it is an eclipsing system. Altogether, therefore, Castor is made up of six suns, four bright and two feeble.

Gamma Crucis is the red star of the Southern Cross. It has been referred to as Gacrux, but this is clearly a made-up name, and is not widely used.

Shaula (Lambda Scorpii). This is the brightest member of the Scorpion's sting, and indeed the name comes from Al Shaulah, the Sting. The other stars in the pattern are Kappa or Girtab (magnitude 2.41), Upsilon or Lesath (2.69), G (3.21) and Q (4.29). Shaula is a normal B-type star; it makes up a splendid pair with Lesath, and the two give the impression of a very wide naked-eye double. However, there is no real connection; Lesath is 1560 light-years away, so that it is much more remote than Shaula. It is also more powerful, and is 16,000 times as luminous as the Sun. From much of Europe the Sting is always very low down, and from England the southernmost part of Scorpius does not rise at all.

Bellatrix (Gamma Orionis). One of the stars in the main pattern of the Hunter. The name means "female warrior"; astrologically it was associated with military prowess. To the natives of the Amazon River, it was the Young Boy in a Canoe. It is the least luminous of the main stars of Orion, and is associated with diffuse nebulosity.

Al Nath (Beta Tauri; formerly Gamma Aurigae). The name comes from Al Natih, the Butting One, since it lies on one of the horns of the Bull. To the Hindus it represented Agni, the God of Fire. It is a normal B-type star; it lies close to the galactic anti-centre, that is to say, the point in the sky exactly opposite to the position of the centre of the Galaxy, which is hidden behind the glorious star-clouds of Sagittarius.

Miaplacidus (Beta Carinae). Originally known as Beta Argûs. In 1680 Edmond Halley tried to introduce a new constellation, Robur Caroli (Charles' Oak) with Miaplacidus as its chief star, but the constellation was never accepted, and swiftly disappeared from the maps. Miaplacidus makes up a triangle with Canopus and the stars of the False Cross.

Alnilam (Epsilon Orionis). The central and brightest star of Orion's Belt; the others are Zeta or *Alnitak* (from Al Nitak, the Girdle) and Delta or Mintaka. All three are very hot and luminous. Alnitak is a fine double; the components are of magnitudes 1.9 and 4.0, with a separation of 2.4 seconds of arc. This is a binary, with a period of 1509 years, and there is a third star, of magnitude 9.9, at 58 seconds of arc which is not a true member of the group.

Alnair (Alpha Gruis). This is the leader of the southern constellation of Grus, the Crane, which is quite distinctive; two of its fainter stars, Delta and Mu, are very wide naked-eye optical doubles. Origi-

The Electromagnetic Spectrum

Visible light makes up only a very small part of the total range of wavelengths, or electromagnetic spectrum. Light is a wave motion, and the colour of the light depends upon the wavelength; red has the longest wavelength, violet the shortest. If the wavelength is greater than red, or shorter than violet, we cannot see the radiation, though we can detect it in other ways.

Short-wave radiation is measured in units known as Ångströms or nanometres. The Ångström, named in honour of the 19th-century Swedish physicist Anders Ångström, is equal to one ten-thousand millionth of a metre; one nanometre is one thousand-millionth of a metre, so that for instance 1 nm+10 A.

The accepted divisions are as follows:

Below 0.01 nm:	Gamma-rays.
0.01–10 nm:	X-rays.
10–400 nm:	Ultra-violet. (EUV, Extreme Ultra-Violet, extends from 10 to 120 nm.)
Visible light:	400–700 nm (+4000–7000 A.)
Intra-red:	700 nm–1 mm.
Sub-millimetre:	0.3 mm–1 mm.
Microwaves:	1mm–0.3 metres.
Radio waves:	Over 0.3 metres.

nally Grus was partly included in Piscis Australis, and the name Alnair comes from Al Na'ir, the Bright One of the Fish's Mouth. It is a white star; there is a good colour contrast with Al Dhanab or Beta Gruis, the second star of the Crane (magnitude 2.11) which is a beautiful warm orange hue.

Alioth (Epsilon Ursae Majoris). The brightest of the seven stars of the Plough or Dipper pattern of the Great Bear. It is unusual inasmuch as its spectrum shows strong lines of the elements chromium and europium, and it seems to have a variable magnetic field. Next in order of brightness in the Plough come Eta (*Alkaid*, otherwise known as Benetnasch) and Alpha (*Dubhe*, from Thahr al Dubb al Akbar, the Back of the Great Bear). The pattern is completed by the fainter Beta (Merak, magnitude 2.37); Gamma (Phad or Phekda, 2.44); Delta (Megrez, 3.31) and the famous double Zeta (Mizar, 2.09). Dubhe is orange, the others white. Five of the seven stars share a common motion in space, and form a moving cluster, but Dubhe and Alkaid are moving in the opposite direction, so that eventually the Plough pattern will become distorted. The celebrated Whirlpool Galaxy, M.51, lies close to Alkaid, just across the border of the neighbouring constellation of Canes Venatici (the Hunting Dogs). M.51 was the first spiral to be recognized as such, by Lord Rosse in 1845, using his home-made 72-inch reflector at Birr Castle in Ireland.

Regor (Gamma Velorum; originally Gamma Argûs). With its declination of -47 degrees, Regor is unavailable from any part of Europe. It is a very hot, unstable star of spectral type W, showing both bright and dark lines in its spectrum; stars of this type are associated with expanding gas-shells, with material streaming out quickly all the time. Regor seems to have a mass 15 times that of the Sun, and a diameter 17 times as great as that of the Sun. It is a very wide, easy double; the components are of magnitudes 1.9 and 4.2, and are separated by over 41 seconds of arc. There are two faint optical companions at 62 and 93 seconds of arc respectively.

Mirphak (Alpha Persei). The leader of Perseus, the mythological hero who killed Medusa the Gorgon, a hideous creature with a woman's head and hair made of snakes. It adjoins Andromeda, and is a normal star of type F. An old name for it was Algenib, but this has been dropped, as the star Gamma Pegasi in the Square is now known as Algenib. The second star in Perseus is Algol, the prototype eclipsing binary, which is normally of magnitude 2.1, but falls to magnitude 3.4 every 2.9 days when the brighter component is partially hidden by the secondary star.

Kaus Australis (Epsilon Sagittarii). Sagittarius, the Archer, is the southernmost of the Zodiacal constellations. It has sometimes been linked with Chiron, the wise Centaur who was tutor to the Argonauts and other heroes; another idea is that Chiron invented the constellation to act as a guide to travellers, but all in all it seems much more likely that it is named in honour of Crotus, the satyr or centaur who was the son of the goat-god Pan. Though it has no first-magnitude star, it is rich and important. Kaus Australis is the "Southern Bow"; the central bow is Delta (Kaus Meridionalis, magnitude 3.01) and

The Structure of the Atom

All matter is made up of atoms, and these atoms are almost inconceivably small. The classical picture of an atom was given by the Danish scientist Niels Bohr in the early 20th century. There is a central nucleus, carrying a positive charge of electricity; round this move "planetary" electrons, each of which carries unit negative charge. The nature of the atom depends upon the number of planetary electrons. Hydrogen has one electron; helium 2; lithium 3, and so on up to uranium, which has 92. There is a complete sequence, and we know that there are no missing elements, because one cannot have half an electron. Elements with more than 92 electrons have been made in our laboratories, but are unstable, and have not been found occurring naturally.

The number of planetary electrons in a complete atom is always just sufficient to balance the positive charge of the nucleus, so that the whole atom is electrically neutral. If, however, an atom is robbed of one or more of its electrons, it is said to be ionized. The nucleus is then left with a positive charge, and there are free or unbound electrons.

Rejected Constellations

Halley's Robur Caroli is only one of many constellations which have been introduced and subsequently abandoned. Among them are:

Lilium:	The Lily.
Triangulum Minor:	The Little Triangle.
Tarandus:	The Reindeer.
Solitarus:	The Solitaire.
Taurus Poniatowski:	Poniatowski's Bull.
Psalterum Georgianum:	George's Lute.
Spectrum Brandenburgicum:	The Sceptre of Brandenburg.
Globus Aerostaticus:	The Balloon.
Officina Typographica:	The Printing Press.
Lochium Funis:	The Log Line.
Noctua:	The Night Owl.
Felis:	The Cat.

Most of these are no loss, though I admit to a feeling of regret that we have summarily rejected the Owl and the Pussycat!

the Northern Bow is Lambda (Kaus Borealis, 2.81). Epsilon is a B-type star, lying in the Milky Way; there is a 7th-magnitude optical companion at a separation of over 3 minutes of arc.

In Sagittarius the sequence of Greek letters is very much out of order. Following Epsilon, the brightest stars are Sigma (2.02) and Zeta (2.59). Alpha or Rukbat and Beta or Arkab are only just above the fourth magnitude.

Wezea (Delta Canis Majoris), one of the triangle with Adhara and Aludra. It is very remote — and it is more than 5000 times as luminous as Sirius.

Avior (Epsilon Carinae, originally Epsilon Argûs). The red member of the False Cross. The colour contrast with the other members of the Cross (Iota Carinae, and Delta and Kappa Velorum) is striking, particularly when viewed with binoculars.

Sargas (Theta Scorpii). The southernmost of the really bright stars in the Scorpion. It is a very luminous F-type supergiant.

Menkarlina (Beta Aurigae). The name comes from Al Mankib Adhi'l Inan, the Shoulder of the Rein-holder; one of the four stars making up the quadrilateral of Auriga — the other three are Capella, Theta Aurigae (magnitude 2.6) and Iota or Hassaleh (2.7). It is an eclipsing binary with a very small magnitude range and a period of 3.96 days. The two components are almost equal, and move in circular orbits, or nearly so. About one-quarter of the diameter of each star is occulted at mid-eclipse. Both are of type A, and are no more than about 8,000,000 miles (12,000,000 km) apart. There is an optical companion, of below magnitude 10, at a separation of 184 seconds of arc.

Alpha Trianguli Australe (Atria; a clearly made-up name, not often used). The Southern Triangle is one of the few constellations to resemble the object after which it is named, as the three leaders, Alpha, Beta (2.85) and Gamma (2.89), really do form a triangle, close to the Pointers (Alpha and Beta Centauri). Alpha is an orange K-type star whose colour is very evident through binoculars.

Alhena (Gamma Geminorum). The third star of Gemini; it lies in the Milky Way, between Pollux and Betelgeux. The name comes from the Arabic Al Han'ah, the Foot of the Twins. It forms a triangle with Betelgeux and Procyon.

Alpha Pavonis. Pavo, the Peacock, is one of the four Southern Birds; the others are Grus (the Crane), Phoenix (the Phoenix) and Tucana (the Toucan). Alpha, which has never been dignified with a proper name is rather isolated from the rest of the constellation; it is easy to identify, as Alpha Centauri and Alpha Trianguli Australe show the way to it. It is a B-type star, 700 times as luminous as the Sun.

Delta Velorum (the strange name of Koo She is seldom used). A white A-type star, one of the four making up the False Cross. It was originally known as Delta Argûs.

Alphard (Alpha Hydrae). Because of its very isolated position in the sky, Alphard is known as the Solitary One, though Tycho Brahe called it Cor Hydrae (the Watersnake's Heart). The best way to identify it is to use Castor and Pollux as pointers. It is obviously orange, with a K-type spectrum. Sir John Herschel was convinced that it was variable; this has not been confirmed, and it is an awkward star to estimate, because of the signal lack of suitable comparisons. If it does vary, the magnitude range is very small. There is a 9.5-magnitude optical companion at a separation of 283 seconds of arc (position angle 153 degrees).

Mirzam (Beta Canis Majoris). One of the Adhara triangle, it is a pulsating variable with a very small range (0.03 of magnitude). Stars of this type are known either as Beta Canis Majoris or Beta Cephei variables; another member of the class is Beta Crucis. All are very luminous B-type giants.

Algieba (Gamma Leonis). This is a member of the Sickle, and has been described with Regulus. It is a fine double.

Polaris (Alpha Ursae Minoris). The northern Pole Star has been of immense value to navigators; it is a pity that there is no southern counterpart. An old name for it was Cynosura. It is very slightly variable, and pulsates in a period of 4 days, but there is evidence that the amplitude has decreased during the past few years, and it has even been suggested that in the near future the pulsations will stop altogether; we must wait and see — if they do cease, we will have witnessed a real step in the star's evolution. It is a notable double; the companion is of magnitude 9, and the separation is 18.4

Trails of stars near the north celestial pole.
Photograph: Patrick Moore, 1995.

seconds of arc (position angle 218 degrees). The companion was discovered by Sir William Herschel in 1780, and is an easy object when viewed through a telescope of 2.5 inches aperture or more. The two share a common motion in space, and are certainly associated, but they are a long way apart, and the revolution period must amount to many thousands of years.

It is always interesting to take a time-exposure of the region centred on Polaris; the trails are very evident. Polaris itself leaves a short trail, because it is not exactly at the pole. In fact it will be at the

Telluric Lines

Mintaka, the northernmost star of Orion's Belt, lies very close to the celestial equator. In 1834 Sir John Herschel announced that it was variable, but it is now known to be an eclipsing binary with so small a range that Herschel's observation may not have been valid.

As Mintaka is a spectroscopic binary, its lines show displacement by the usual Doppler effect, and the star is receding from us at 12 miles per second (20 km per second). However, in 1904 J.F. Hartmann, at the Potsdam Observatory, found that some of the absorption lines in the spectrum did not share in this displacement. Therefore they could not originate in the star: they are due to the absorbing effects of gases in our own atmosphere, and are now known as telluric lines (Tellus, the Earth).

The Legend of Aries

Aries represents a ram sent by Mercury or Hermes, messenger of the gods, to rescue the children of the King of Thebes from assassination by their archetypal wicked step-mother. The ram had a golden fleece, and possessed the remarkable ability to fly. The ram carried out its mission; sadly the girl (Helle) fell off and met her death in that part of the sea still called the Hellespont, but the boy (Phryxus) survived, and arrived safely in the kingdom of King Aeetes. After the ram's death, its fleece was placed in a sacred grove and was guarded by ferocious dragons, where it stayed until removed by Jason and the Argonauts.

closest to the pole on 24 March 2100, when its declination will be +89° 32' 51"; after that the pole will start to move away from it once more.

Hamal (Alpha Arietis). Aries used to be the first constellation of the Zodiac, though the vernal equinox has now moved into Pisces because of the effects of precession. Aries represents the ram with the golden fleece, captured by the Argonauts led by Jason; the name Hamal means "the Head of the Sheep". The constellation is marked by a group of three stars not far from Andromeda; the others are Beta or Sheratan (magnitude 2.64) and Gamma or Mesartim, which is a fine double with equal components, both of magnitude 4.6. Hamal itself has a K-type spectrum, and binoculars show that it is distinctly orange.

Mira (Omicron Ceti). One of two variable stars which can occasionally exceed the second magnitude. It was first noted in 1596 by David Fabricius, when the magnitude was 3. Subsequently it faded from view; Johann Bayer saw it again in 1603, and allotted it the Greek letter Omicron, but again it disappeared. In 1638 the Dutch observer Phocylides Holwarda recovered it, and realized that it appeared and vanished periodically; it was the first known variable star, and the Danzig astronomer Hevelius christened it Mira, "the Wonderful". It has given its name to a whole class of long-period variables.

Mira has a period of 332 days, though this is not quite constant. The maximum may range from magnitude 1.7 to below 4; there are vague accounts that it was once known to rise to 1.2, but certainly this is most unusual, and Mira is a naked-eye object for only a few weeks in every year. The minimum magnitude is just below 10. There is a sub-dwarf companion of magnitude 9.5, which is itself variable and has a variable star designation (VZ Ceti). The separation is only 0.3 of a second of arc. There is no doubt that the two stars are associated, but again they are a long way apart.

Gamma Cassiopeiae. The centre star of the famous W of Cassiopeia in the far north of the sky. It is an unstable star, and periodically ejects shells of gas. Generally it is of about magnitude 2.2, but in 1937 it rose to 1.6; subsequently it fell to just below magnitude 3 before returning to "normal". It is always worth watching, since it seems to be quite irregular in behaviour. In the W, Beta or Chaph (2.27) and Delta or Ruchbah (2.68) make useful comparisons; Alpha or Shedir, a reddish K-type star, is officially ranked of magnitude 2.23, but is probably variable to the extent of a few tenths of a magnitude.

MAGNITUDES 2 TO 2.5

This completes our survey of stars down to the second magnitude. Let us end by listing the remaining stars down to magnitude 2.5:

Star		Apparent magnitude	Absolute magnitude	Spectrum	Distance lt-yrs
Sigma Sagittarii	Nunki	2.02	-2.0	B3	204
Beta Ceti	Diphda	2.04	0.2	K0	60
Beta Andromedae	Mirach	2.06	-0.4	M0	88
Alpha Andromedae	Alpheratz	2.06	-0.1	A0	72
Theta Centauri	Haratan	2.06	1.7	K0	46
Kappa Orionis	Saiph	2.06	-6.9	B0.5	2100
Beta Ursae Minoris	Kocab	2.08	-0.3	K4	95
Alpha Ophiuchi	Rasalhague	2.08	0.3	A5	62
Zeta Ursae Majoris	Mizar	2.09	0.4+2.1	A2+A6	59
Beta Gruis	Al Dhanab	2.11	2.4	M3	173
Beta Persei	Algol	2.12v-	0.2	B8	95
Gamma Andromedae	Algieba	2.14	-0.1	K0+A0	121
Beta Leonis	Denebola	2.14	1.7	A3	39
Gamma Cygni	Sadr	2.20	-4.6	F8	750
Lambda Velorum	Al Sahail al Wazn	2.21	-4.4	K5	490
Alpha Coronae Borealis	Alphekka	2.23	0.6	A0	78
Alpha Cassiopeiae	Shedir	2.23v?	-0.9	K0	120
Zeta Puppis	Suhail Hadar	2.25	-7.1	O5.8	2400
Delta Orionis	Mintaka	2.23v	-6.1	O9.5	2350
Gamma Draconis	Eltamin	2.25	-0.3	K5	101
Iota Carinae	Tureis	2.25	-4.7	F0	800
Epsilon Scorpii	Wei	2.29	-0.1	K2	65
Alpha Lupi	Men	2.30	-4.4	B1	6800
Epsilon Centauri		2.30	-3.5	B1	490
Eta Centauri		2.31	-2.9	B3	110
Delta Scorpii	Dschubba	2.32	-4.1	B0	550
Beta Ursae Majoris	Merak	2.37	1.2	A1	62
Epsilon Boötis	Izar	2.38	-0.9	K0	150
Epsilon Pegasi	Enif	2.38	-4.4	K2	520
Alpha Phoenicis	Ankaa	2.39	0.2	K0	78
Beta Pegasi	Scheat	2.4v	-1.4	M2	176
Kappa Scorpii	Girtab	2.41	-3.0	B2	390
Gamma Ursae Majoris	Phad	2.44	0.6	A0	75
Eta Canis Majoris	Aludra	2.44	-7.0	B5	2500
Alpha Cephei	Alderamin	2.44	1.9	A7	46
Epsilon Cygni	Gienah	2.46	0.2	K0	81
Alpha Pegasi	Markab	2.49	0.2	B9	100
Kappa Velorum	Markeb	2.50	3.0	B2	390

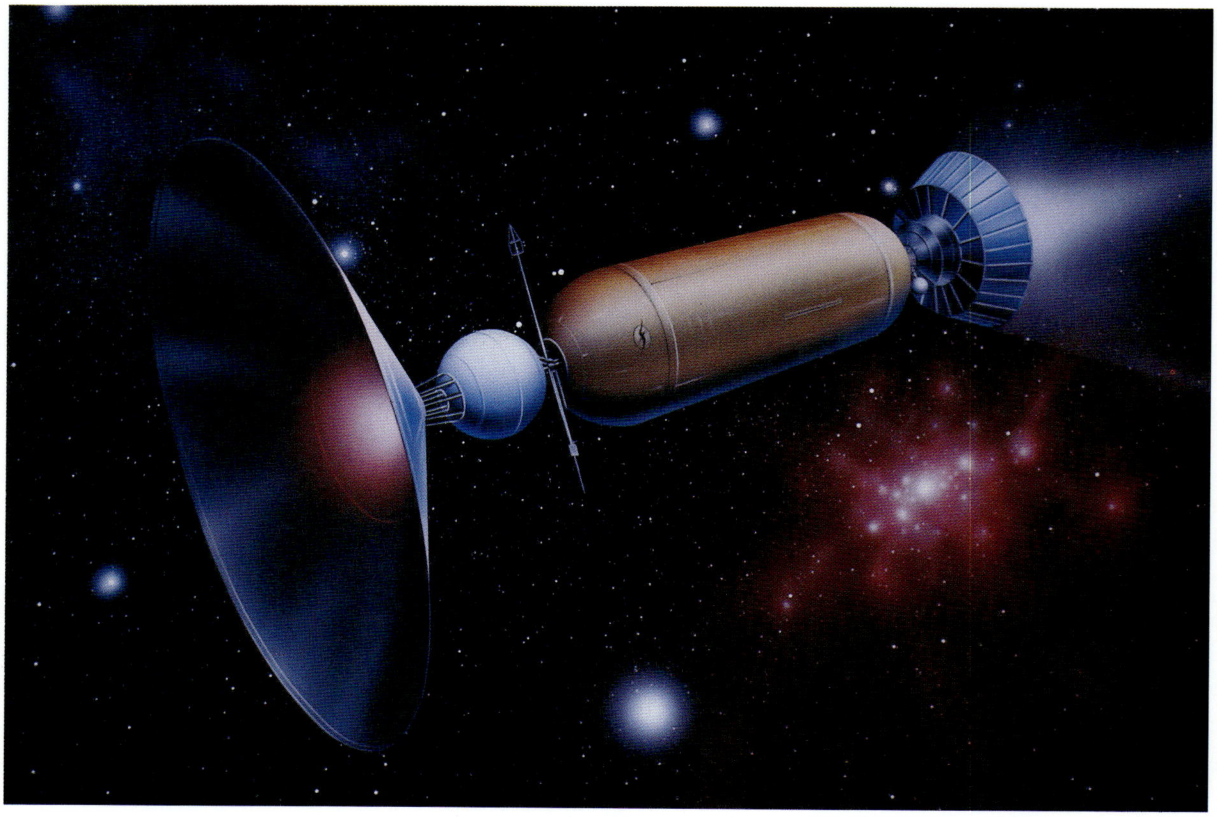

Space travel of the future? An artists's impression of a ramjet craft.
Paul Doherty

GLOSSARY

ABSOLUTE MAGNITUDE. The *apparent* magnitude which a star would have if it were observed from a standard distance of 10 *parsecs* (32.6 *light-years*).

ALTITUDE. The angular distance of a celestial body above the horizon.

ÅNGSTRÖM UNIT. One hundred-millionth part of a centimetre.

APHELION. The point in its orbit when a planet or comet is at its greatest distance from the Sun.

APPARENT MAGNITUDE. The apparent brightness of a celestial body; the lower the magnitude, the brighter the object. The magnitude of the Sun is about -27; the Pole Star, +2; normal limit of naked-eye visibility, +6; faintest detectable with modern equipment around +29.

ASTRONOMICAL UNIT. The mean Earth-Sun distance; approximately 93,000,000 miles (149,598,000 km).

BINARY STAR. A star made up of two components, moving round their common centre of gravity.

BLACK HOLE. A region round an old, collapsed star from which not even light can escape.

CELESTIAL SPHERE. An imaginary sphere surrounding the Earth, whose centre is coincident with the centre of the Earth.

CEPHEID. A short-period variable star; the variations are very regular. Their real luminosities are linked with their periods; the longer the period, the more luminous the star. The name comes from the prototype star, Delta Cephei.

CIRCUMPOLAR STAR. A star which never sets. Thus Ursa Major is circumpolar from England, while the Southern Cross is circumpolar from New Zealand.

CORONA. The outer part of the Sun's atmosphere.

COSMOLOGY. The study of the universe considered as a whole.

CULMINATION. The maximum altitude of a celestial body above the horizon.

DAY, SIDEREAL. The mean interval between successive *culmination* of the same star: 23h 56m 4s.091.

DECLINATION. The angular distance of a celestial body north or south of the celestial equator. It corresponds to latitude on Earth.

DIRECT MOTION. Movement of revolution or rotation in the same sense as that of the Earth.

DOPPLER EFFECT. The apparent change in wavelength of a light from a luminous body which is in motion relative to the observer. With an approaching object the wavelength is effectively shortened, the object appears "too blue"; with a receding object the wavelength is lengthened, and the object appears "too red". This affects the positions of the lines in a stellar spectrum; the amount of blue or red shift gives a key to the star's velocity.

DOUBLE STAR. A star made up of two (or more) components, either genuinely associated (binary pair) or merely lined up (optical pair).

ECLIPSING BINARY (or ECLIPSING VARIABLE). A *binary* system in which one component is regularly *occulted* by the other, so causing a decrease in the *apparent magnitude* of the system.

ECLIPTIC. The projection of the Earth's orbit on to the *celestial sphere*. It may also be defined as the apparent yearly path of the Sun against the stars.

EQUATOR, CELESTIAL. The projection of the Earth's equator on to the *celestial sphere*.

EQUINOXES. The two points at which the *ecliptic* crosses the *celestial equator*; crossing of the Sun from south to north (Vernal Equinox or First Point of Aries, around March 21 each year) and from north to south (First Point of Libra, around September 22 each year). Owing to *precession*, the vernal equinox is now in Pisces and the autumnal equinox in Virgo.

ESCAPE VELOCITY. The minimum velocity which a body must have if it is to escape from a planet, or other body, without being given extra impetus. The Earth's escape velocity is 7 miles (11 km) per second.

EXTINCTION. The apparent dimming of a celestial body when low over the horizon, so that its light comes to us via a denser layer of the Earth's atmosphere.

FLARES. Brilliant eruptions in the outer part of the Sun or any other star.

FRAUNHOFER LINES. The dark absorption lines in the spectrum of the Sun.

GALAXIES. Systems made up of stars, *nebulae* and interstellar matter.

GALAXY, THE. The Galaxy which contains our Sun. It includes about 100,000 million stars.

GAMMA-RAYS. Radiations of extremely short wavelength.

GLOBULES. Dark patches inside *nebulae*; they are probably embryo stars.

H.I AND H.II REGIONS. Clouds of hydrogen in the *Galaxy*; H.I if the atoms are neutral, H.II if they are ionized.

HALO, GALACTIC. The spherical-shaped cloud of stars round the main Galaxy. Globular clusters are halo objects.

HELIACAL RISING. The rising of a star or planet at the same time as the Sun, though the term is generally used to denote the time when the object is first visible in the dawn sky.

HERTZSPRUNG-RUSSELL (H-R) DIAGRAM. A diagram in which

the stars are plotted according to their spectral type and *absolute magnitude*.
HORIZON. The great circle on the *celestial sphere* which is everywhere 90 degrees from the observer's *zenith*.
HUBBLE CONSTANT. The rate of increase in the recessional velocity of a galaxy with increased distance from Earth. Its current value is uncertain.
INTERFEROMETER, STELLAR. An instrument for measuring star diameters. It is based on the phenomenon of the interference of light-waves.
ION. An atom which has lost one or more of its planetary electrons, so that it has a positive charge of electricity.
KILOPARSEC. One thousand *parsecs*.
LIGHT-YEAR. The distance travelled by light in one year; 5,880,000,000,000 miles (nearly 6 million million miles) or 9.4607 million million km.
LOCAL GROUP. A cluster of more than two dozen galaxies, one of which is our Galaxy.
MAIN SEQUENCE. A band in the H-R Diagram, from top left to bottom right, including most normal stars apart from giants.
MEGAPARSEC. One million *parsecs*.
NEBULA. A cloud of gas and dust in space. (Galaxies were known as extragalactic nebulae before their true nature was realized.)
NEUTRINO. A fundamental particle with no electrical charge and little rest mass (perhaps none at all).
NEUTRON. A fundamental particle with no electrical charge, but a mass virtually equal to that of a *proton*.
NEUTRON STAR. The remnant of a very massive star which has exploded as a *supernova*, and left a core composed only of *neutrons*.
NOVA. A star which suddenly flares up to many times its normal brilliancy, remaining bright for some time before fading back to normal. All novae seem to be *binary* systems.
OBLIQUITY OF THE ECLIPTIC. The angle between the *ecliptic* and the *celestial equator*: 23° 26' 45".
OCCULTATION. The covering-up of one celestial body by another.
OPPOSITION. The position of a planet when exactly opposite to the Sun in the sky.
ORBIT. The path of a celestial object.
PARALLAX, TRIGONOMETRICAL. The apparent shift in the position of a near object when viewed from two different directions against a distant background.
PARSEC. The distance at which a star would have a *parallax* of one second of arc (hence the name; *par sec*). It is equal to 3.26 *light-years*, or 206,265 *astronomical units*. Apart from the Sun, no star is as close as this.

PERIHELION. The point in its orbit when a planet or comet is nearest to the Sun.
PHOTOMETER. An instrument used to measure the intensity of a light-source.
PHOTON. The smallest 'unit' of light.
PLANETARY NEBULA. A small, dense, hot star surrounded by a shell of gas.
POLES, CELESTIAL. The north and south points of the *celestial sphere*.
POPULATIONS, STELLAR. Two main types of star regions; I (in which the brightest stars are hot and white) and II (in which the brightest stars are old red giants).
POSITION ANGLE. The apparent direction of one object with reference to another, measured from the north point of the main object through east, south and west.
PRECESSION. The apparent slow movement of the celestial poles, due to the pull of the Sun and Moon on the Earth's equatorial bulge. It also means a shift in the position of the celestial equator, and hence of the equinoxes; the vernal equinox shifts by 50 seconds of arc per year, and has moved from Aries into Pisces.
PROPER MOTION, STELLAR. The individual movement of a star on the *celestial sphere.*
PROTON. A fundamental particle with unit positive electrical charge.
QUANTUM. The amount of energy possessed by one *photon* of light.
QUASAR. A very remote, superluminous object, now believed to be the core of a very active galaxy.
RADIAL VELOCITY. The movement of a celestial body toward or away from the observer; positive if receding, negative if approaching.
RADIANT. The position in the sky from which the meteors of a particular shower appear to radiate.
RETROGRADE MOTION. Orbital or rotational movement in the sense opposite to that of the Earth's motion.
RIGHT ASCENSION. The angular distance of a celestial body from the vernal equinox, measured eastward. It is usually measured in units of time so that it is equal to the time-difference between the culmination of the vernal equinox and the culmination of the body concerned.
SCINTILLATION. Twinkling. It is due entirely to the effects of the Earth's atmosphere.
SEYFERT GALAXIES. Galaxies with relatively small, bright nuclei and weak spiral arms. Some of them are strong radio emitters.
SHOOTING-STAR. The luminous trail left by a meteor as it plunges through the Earth's upper air, and is excited to incandescence. Meteors are cometary debris, and burn out well before reaching ground level; they are not associated with meteorites.
SIDEREAL TIME. Time by the stars instead of the Sun. The local

time reckoned according to the apparent rotation of the *celestial sphere*. When the vernal equinox reaches culmination, the sidereal time is 0 hours.
SOLSTICES. The position of the Sun when at its maximum *declination* of approximately 23½ degrees; around June 22 (northern) and 22 December (southern). The northern solstice lies in Gemini, the southern in Sagittarius.
SPECTROSCOPIC BINARY. A binary system whose components are too close together to be separated visually, but which can be detected by the shifts in their spectral lines.
SUPERNOVA. A colossal stellar outburst; due to the total destruction of a white dwarf star (Type I) in a binary system or the collapse and subsequent explosion of a very massive star (Type II).
TRANSIT. (1) The passage of a celestial body across the observer's meridian. (2) The projection of one celestial body against another.
TWILIGHT. The state of illumination after sunset, or before sunrise, when the Sun is less than 18 degrees below the horizon.
VARIABLE STARS. Stars which change in brilliancy over definite periods. Secular variables are stars which are alleged to have shown permanent change in historical times.
WOLF-RAYET STARS. Very hot, unstable stars which have expanding shells of gas, and have bright lines in their spectra.
ZENITH. The observer's overhead point (altitude 90 degrees).
ZENITH DISTANCE. The angular distance of a celestial object from the *zenith*.
ZODIAC. A belt stretching round the sky, 8 degrees to either side of the *ecliptic*, in which the Sun, Moon and principal planets are always to be found.
ZODIACAL LIGHT. A cone of light rising from the horizon and extending along the *ecliptic*, visible only when the Sun is a little way below the horizon. It is due to thinly-spread material near the main plane of the Solar System

INDEX